OPEN TUBULAR COLUMNS
IN GAS CHROMATOGRAPHY

OPEN TUBULAR
COLUMNS
IN GAS CHROMATOGRAPHY

L. S. Ettre

Chief Applications Chemist,
Analytical Division
The Perkin-Elmer Corporation, Norwalk, Connecticut

With a Foreword by
M. J. E. Golay

PLENUM PRESS
NEW YORK
1965

Library of Congress Catalog Card Number 65-13583

©1965 Plenum Press
A Division of Consultants Bureau Enterprises, Inc.
227 West 17th Street • New York, N. Y. 10011

Printed in the United States of America

To My Colleagues
at
Perkin-Elmer

Foreword

For my past sins, Leslie Ettre has given me the privilege of writing a few words to preface his excellent little book. It gives me great pleasure to do so, because of the many years of fruitful collaboration we have had at Perkin–Elmer, because it is refreshing to see a treatise in gas chromatography in which the theoretical treatment has been bared to its essentials, without a mushrooming of formulae which, by means of an ever increasing number of parameters, account for more and more, and explain less and less, and because the author has recognized that the gas chromatographic column is a nearly passive element in its own right which deserves to have a treatise written nearly exclusively about it, just as electrical circuit theory can be discussed without elaborate references to vacuum tubes and meters.

I wish this conscientiously written volume the success it deserves.

M. J. E. GOLAY

Preface

Gas chromatography is a separation technique used primarily in analytical chemistry. Therefore, it is evident that special emphasis should be placed on that particular part of the apparatus in which the separation takes place. This part is the column, the heart of the gas chromatograph.

The goal of researchers in the field of gas chromatography has been—from the beginning—to understand the separation process so that they might design columns with the best possible performance. Such investigation led M. J. E. Golay in 1956–1957 to the development of a special column type. These columns are not filled with the ordinary column material, which consists of a solid support and a stationary (liquid) phase; rather the inside column wall itself serves as the support and is coated with a thin film of the stationary phase. As pointed out by W. W. Brandt [24], this invention "ranks as one of the two or three major original developments in gas chromatography since its discovery by James and Martin."

In the years since Golay's invention, these columns have found wide application in practically every field where gas chromatography is used. The publications on the theory, instrumentation, and applications of these columns are voluminous. However, with the exception of Kaiser's excellent book [160], originally written in 1961, the accumulated knowledge has not been summarized in one single publication.

In November 1963, the author had the honor to present a lecture at the Eastern Analytical Symposium in New York City, discussing in detail some theoretical and practical aspects of open tubular columns and their applications. In the following months, this lecture was repeated by invitation—with the content continuously updated—on six different occasions, including the yearly Gas Chromatography Institutes held at Canisius College, Buffalo, N.Y., and Fisk University, Nashville, Tenn. In the discussions following these presentations, it was often suggested that the text of the lecture be made available in printed form. This is how the idea for this book was born: it is based on these lectures, but the original text has been significantly enlarged and modified; furthermore, chapters discussing questions not included in the oral presentations were added in order to cover the subject completely.

This book was written for the *practical gas chromatographer*: its principal aim is to help him better understand this exceptional tool. After a brief introduction, some theoretical aspects are discussed in order to allow certain conclusions necessary in the actual application of these columns to be drawn. This treatment was felt to be most profitable since theory and practice cannot be separated. Subsequently, questions concerning the preparation of the various column types and special problems in connection with the gas chromatographic system in which these columns are used are discussed. Finally, a complete bibliography of publications dealing with the theory and applications of open tubular columns in gas chromatography, with a short subject index, is given. Bracketed numbers in the text refer to this bibliography, while other literature sources are given as footnotes.

Most of the chromatograms shown in the figures were selected to illustrate certain specific points discussed in the text rather than to show the analysis of typical samples, primarily because it is the author's experience from his long activity in gas chromatography that no "typical sample" exists. Also, the principal purpose of the chromatograms is to illustrate some practical applications, to draw attention to the most important fields to which open tubular columns can be successfully applied, and not to reproduce the best chromatograms ever published. It was felt that the basic goal of the book is not to give a collection of chromatograms but to help in better understanding the characteristics of open tubular columns.

The various units of measurement used by different authors in describing the columns and analytical conditions represented a serious problem. Although in this book the metric system is used whenever possible, data on column length and inlet pressure were left as given originally by the authors because the transformation of feet and pounds per square inch into the corresponding metric units would often result in irrational numbers. However, the internal diameter of the column is always given in metric units, in order to make comparison and calculation simpler.

Since this book discusses the open tubular columns, the instrumentation used to obtain the various chromatograms is not mentioned in the figure captions; however, it is summarized in the Supplement.

Great care was taken to establish a logical nomenclature. The symbols used are based on the recommendations of the various international symposia and the special I.U.P.A.C. Committee; symbols corresponding to terms not included in their treatment were established by selecting the most logical usage. In this respect, our goal was to have only *one* meaning for each symbol.

This book is based on both the literature and the experiences collected in the last eight years at the various laboratories of The Perkin–Elmer Corporation, where the original work of Dr. Golay was carried out. The author wishes to express his sincere appreciation to all colleagues and sources listed in the following pages who placed at his disposal various published and unpublished material. Special thanks are due to W. Averill, N. Brenner, R. D. Condon, and S. D. Norem at Perkin–Elmer and Dr. C. Horváth at the Research Laboratory of the Massachusetts General Hospital, Boston, who read parts of the manuscript and were instrumental, by their thoughtful advice and criticism, in molding its final shape; further, to the management of The Perkin–Elmer Corporation, Norwalk, Conn., for permission to write this book and for allowing the use of various company facilities. The fine cooperation of J. R. Piccolo and E. Hagenburger in the preparation of the illustrations and of Mrs. E. Barrett in typing the manuscript is particularly noteworthy. Last but not least, the efforts of Plenum Press in making possible the publication of this book in an unusually short time are greaty appreciated.

It is the author's sincere hope that this work will give the practical gas chromatographer a better understanding of the characteristics of open tubular columns and help him in improving his everyday work.

Norwalk, Connecticut   L. S. ETTRE
August 20, 1964

Acknowledgments

The illustrations in this book, aside from those constructed by the author, come from a variety of sources. Reproduction of the following illustrations is authorized through the courtesy of the publications, corporations, and persons named:

Analytical Chimica Acta (Elsevier Publishing Co., Amsterdam, The Netherlands): Figure 30.

Analytical Chemistry (American Chemical Society, Washington, D.C., USA): Figures 27, 38, 50, 51.

Brennstoff-Chemie (Girardet Verlag, Essen, Germany): Figures 24, 25.

Chemistry & Industry (Society of Chemical Industry, London, England): Figure 31.

Instrument News (The Perkin–Elmer Corp., Norwalk, Conn., USA): Figures 16, 17.

Journal of Chromatography (Elsevier Publishing Co., Amsterdam, The Netherlands): Figures 28, 61, 74, 75.

Journal of Gas Chromatography (Preston Technical Abstracts Co., Evanston, Ill., USA): Figures 5, 7, 34, 37, 42, 43, 44, 48, 49, 54, 73.

Nature (Macmillan (Journals) Ltd., London, England): Figures 56, 57, 58, 69.

Barber-Colman Co., Rockford, Ill., USA: Figure 32.

Bodenseewerk Perkin–Elmer & Co., Überlingen, Germany: Figures 18, 26, 29, 36, 45, 47, 65, 67, 76.

Butterworths Inc., Washington, D.C., USA: Figure 53.

Perkin–Elmer Corp., Norwalk, Conn., USA: Figures 6, 19, 20, 21, 35, 46, 55, 59, 64, 66, 68, 70, 71, 72.

W. Averill (Perkin–Elmer Corp., Norwalk, Conn., USA): Figure 41.

E. W. Cieplinski (Perkin–Elmer Corp., Norwalk, Conn., USA): Figure 66.

C. Horváth (Massachusetts General Hospital, Boston, Mass., USA): Figure 52.

F. J. Kabot (Perkin–Elmer Corp., Norwalk, Conn., USA): Figures 14, 33.

J. Oesterhelt (Bodenseewerk Perkin–Elmer & Co., Überlingen, Germany): Figure 45.

Symbols

a	Constant in Eq. (50)
a_K	Constant in Eq. (51)
a	Accommodation coefficient in gas–solid adsorption chromatography
A	Peak area
b	Constant in Eq. (50)
B	Longitudinal gaseous diffusion term
B_0	Specific permeability
c	Concentration of the coating solution
c_{min}	Minimum detectable concentration in a given sample
C_G	Term of resistance to mass transfer in the gas phase
C_K	Mass transfer term for the kinetics of adsorption and desorption in gas–solid adsorption chromatography
C_L	Term of resistance to mass transfer in the liquid phase
d	Inside diameter of column tubing
d_f	Average thickness of the liquid-phase film
d_p	Effective particle diameter of the support
d_L	Density of the liquid phase
D_G	Gaseous diffusion coefficient
D_L	Liquid diffusion coefficient
f	A square root function of k [see Eq. (29)]
F_1, F_2	Factors used in chapter 2.322
$\bar{F}$	Average carrier gas flow rate
F_a	Carrier gas flow rate measured at column outlet and ambient temperature
F_c	Value of F_a corrected to column temperature
HETP	Height equivalent to one theoretical plate
HETP_{min}	The theoretical minimum value of the HETP curve
j	Carrier gas compressibility correction factor
k	Partition (capacity) ratio
K	Partition coefficient
L	Column length
m	Constant in Eq. (64)
M	Amount of coating in the column
n	Number of theoretical plates
n_c	Number of carbon atoms in a molecule
n_0	Number of theoretical plates calculated from Eq. (47)
n_{req}	Number of theoretical plates required in order to achieve a desired resolution for a given component pair
N	Number of effective plates

OPGV	Optimum practical gas velocity
p_a	Ambient pressure
p_i	Carrier gas inlet pressure
p_o	Carrier gas pressure at column outlet
p_w	Partial pressure of water at ambient temperature
Δp	Pressure drop along the column
PI	Performance index
q	Constant in Eq. (64)
r	Inside radius of column tubing
R	Peak resolution
R^*	Possible best resolution which could be achieved with a certain column for a given component pair
S	Total internal surface area of the column
t_M	Retention time of an inert component (usually air)
t_R	Retention time of the sample component (measured from start)
t'_R	Adjusted retention time of the sample component (measured from the "air peak")
Δt	Distance of two consecutive peak maxima
T_a	Ambient temperature (in °K)
T_c	Column temperature (in °K)
T_0	Initial temperature in programmed-temperature operation
T_F	Final temperature in programmed-temperature operation
$\bar{u}$	Average linear gas velocity
$\bar{u}_m$	Average molecular velocity
u_o	Linear gas velocity measured at column outlet (corrected to column temperature)
$\bar{u}_{\text{opt}}$	Optimum average linear gas velocity
u_s	Linear velocity of the coating solution at column outlet
v_{eff}	Effective volume of one theoretical plate
v_G	Volume of gas phase in one theoretical plate
v_K	Volume of the vaporized sample exclusive of the carrier gas
v_L	Volume of the liquid phase in one theoretical plate
V	Geometrical volume of the column
V_G	Total volume of gas phase in a column
V_L	Total volume of liquid phase in a column
V_s	Necessary volume of coating solution
w_b	Peak width at base (band intercept)
w_h	Peak width at half height
w_i	Peak width at the inflection points
$x_0, ..., x_3$	Distances on the chromatogram used for the calculation of the "air peak" time
α	Relative volatility
α^*	Smallest relative volatility which can be separated with a given resolution on a certain column
β	$= V_G/V_L$
ε	Interparticle porosity
η	Viscosity of the carrier gas
σ	Standard deviation of the Gaussian peak

Contents

First Part

Introduction

1.1 ORIGINS

If the history of scientific developments were to be investigated, it would be found that many are the result of accidental inventions and that the number of significant developments which resulted from logical, theoretical considerations or were based on systematic outlines drawn up *prior* to the actual development is very small. It is interesting to note that both gas–liquid partition chromatography and the application of open tubular columns in gas chromatography are the result of such considerations. Martin and Synge, in 1941, in describing the method of liquid–liquid partition chromatography[1] predicted the possibility of substituting gas for liquid as the mobile phase; similarly, open tubular columns are the result of Golay's theoretical consideration of the behavior of packed columns.

In order to evaluate properly the importance of this new invention, one should look back to the years 1952–1957. Gas chromatography was introduced in 1952 by James and Martin,[2] but the real development started in about 1955 when more and more people became involved in investigations of this new technique. One of the most important questions with which the papers published at that time dealt was the theory of the chromatographic separation process, and the most important publication in this field, by van Deemter, Zuiderweg, and Klinkenberg,[3] describing the basic equation for the performance of a gas chromatographic column appeared in 1956.

[1] A. J. P. Martin and R. L. M. Synge, *Biochem. J.* **35**:1358 (1941).

[2] A. T. James and A. J. P. Martin, *Biochem. J.* **50**:679 (1952).

[3] J. J. van Deèmter, F. J. Zuiderweg, and A. Klinkenberg, *Chem. Eng. Sci.* **5**:271 (1956).

However, this work became really known only in 1957, after Keule-mans' comprehensive book on gas chromatography was published.[4]

Parallel to these studies which had been carried out in Europe, the theory and practice of gas chromatography were also investigated in the United States. In the spring of 1956, Golay presented a paper[5] on a different theoretical approach to this problem. In this paper, based entirely on mathematical considerations, he developed certain equations that yielded a better understanding of the gas chromato-graphic process and drew some conclusions regarding the optimum operation of a gas chromatographic column.

In the months following the presentation of his paper, Golay continued his studies and started to investigate the relationship between the performance of a gas chromatographic (packed) column and such important parameters as the analysis time and the carrier gas pressure which are required for a given performance. In this investigation, a packed column was considered as equivalent to a bundle of capillary tubes coated with the stationary phase and with an inside diameter of the order of the support particles' size. Golay's rough theoretical calculations indicated that the diameter of the capillaries, which determines the resistance to gas flow, should also determine the order of magnitude of the height equivalent to one theoretical plate (HETP).

After checking these theoretical considerations with actual experimental data, an enormous discrepancy was found in that

> "the HETP was ten times greater than the packing granule dimensions while the characteristic dimension controlling the pneumatic resistance to flow was only one-tenth as large."

As Golay himself described later, during this experiment he happened to notice in the laboratory a rather long length of Tygon tubing and, in his own words [109],

> "... out of curiosity, I substituted it for the Celite column, just to see what the air peaks would look like. I was pleasantly surprised to discover that these air peaks were of the correct width predicted by the same rough theory.

---

[4] A. I. M. Keulemans, *Gas Chromatography*, Reinhold, New York, 1957.
[5] M. J. E. Golay, 129th National Am. Chem. Soc. Meeting, Dallas, Texas, April 1956; *Anal. Chem.* **29**:928 (1957).

The next thought was: if the air peaks have the correct width when the column is a length of plain tubing, why not coat the inside wall of this tubing with a retentive substance, and use it for a partition column?''

Golay's first report on his preliminary investigations was dated November 15, 1956. Due to its historical interest, it is reproduced in Figure 1.

November 15, 1956

From   Marcel J. E. Golay

Subject: Progress report of gas chromatographic experimental
         work for September and October 1956

An enormous discrepancy with the theory evolved up to that time was the circumstance that the HETP was ten times greater (0.1 cm) than the packing granule dimensions while the characteristic dimension controlling the pneumatic resistance to flow (0.001 cm) was only one-tenth as large. It had been anticipated that these two quantities, two orders of magnitude apart, would be of the same order of magnitude.

Experiments with open cylindrical columns with an oil coating on the wall were initiated in order to unravel these discrepancies and to obtain a better insight into the separation mechanisms.

An initial experiment with a glass capillary, 1 meter x 0.08 cm, gave inconclusive results because of the near equality of the volumes in the injection block, in the column itself and in the detector (0.8 cm^3). An experiment with an uncoated 10 meter x 0.3 cm length of Tygon tubing gave a $\Delta t/t$ ratio corresponding very closely to the calculated number of theoretical plates of this column for the air peak (about 14,000).

Experiments were then initiated with two steel tubings, 12 feet x 0.055", wall coated with ethylene glycol by filling and wiping with a cotton wad drawn through the tubing. For these experiments, a new detector was constructed in order to reduce its volume.

Figure 1. Facsimile of M. J. E. Golay's first report on his investigations concerning open tubular columns.

A few months after this first internal report, Dr. Golay presented his first paper on the "Theory and Practice of Gas–Liquid Partition Chromatography with Coated Capillaries" [109] at the First Gas Chromatography Symposium of the Instrument Society of America, at East Lansing, Michigan, June 1957.

Golay's work is a classical example of how the proper application of theoretical conclusions can lead to an entirely new invention. This thought was also expressed by A. J. P. Martin, who characterized his paper as "the highlight of the meeting" and the use of open tubular columns as "a method which many of us are going to pursue with considerable enthusiasm from now on."[6]

One year after his first presentation, at the Second International Gas Chromatography Symposium, May 1958, in Amsterdam, The Netherlands, Dr. Golay followed up [110] with a more detailed theory and showed some practical applications of open tubular columns. At the same meeting, Dijkstra and de Goey [68] also demonstrated the possibility of using open tubular columns and described the dynamic coating method used most generally today.

Shortly after these presentations, various groups in different countries started to investigate the potentialities of open tubular columns, and a few months later the publications of Condon [54], Desty et al. [62, 64, 66], Kaiser and Struppe [162], Lipsky et al. [181, 182], Lovelock [187], R. P. W. Scott [243], and Zlatkis et al. [283, 284] gradually confirmed the value of the new concept. From then on, the line of development is unbroken, as more and more researchers and practical gas chromatographers clarified the various aspects of the theory, preparation, and application of these columns.

1.2 NOMENCLATURE

There is considerable confusion regarding the nomenclature of gas chromatographic separation columns built in conformance with Golay's work. They are often called "capillaries," although they are not at all restricted to "capillary" dimensions since "not the smallness but the 'openness' of these columns permits us to realize a two orders of magnitude improvement over the packed columns" [111]. Furthermore, the term "capillary" does not properly characterize this type of column since it has been clearly demonstrated recently[7-9] that *packed* columns of similar diameters as the "capillary capillary

---

[6] A. J. P. Martin, in *Gas Chromatography*, ed. V. J. Coates, H. J. Noebels, and I. S. Fagerson, Academic Press, New York, 1958, pp. 237–247.

[7] I. Halász and E. Heine, *Nature* **194**:971 (1962).

[8] C. Cercy, S. Tistchenko, and F. Botter, *Bull. soc. chim. France* **1962**:2315.

[9] H. V. Carter, *Nature* **197**:684 (1963).

columns" can also be prepared. Thus, if one speaks of "capillary columns," it is not clear which of the two types is meant.

The correct expression for the Golay-type columns is "open tubular columns." This term indicates their most important characteristic—the "openness" of the tubes—and clearly distinguishes them from the packed columns regardless of column diameter, which should be specified separately. In addition, this term allows us to indicate in both cases the type of separation process involved. In the case of *partition* columns, the stationary phase is either held by a packed bed of porous support material (packed columns) or the column wall itself (open tubular columns); on the other hand, in the case of *adsorption* columns, the adsorbent can either completely fill the column (packed columns) or be disposed in a thin layer on the inside wall, leaving an open, unrestricted path through the column (open tubular columns).

In defining the nomenclature, one more distinction is necessary. At present, three types of open tubular columns are being described in the literature. The first two types are partition columns, i.e., the inside of the column is coated with the stationary (liquid) phase: in "classical" open tubular columns the liquid phase is deposited directly on the inside column wall, but recently columns have been developed in which the inside surface of the tubing is enlarged by having a thin, porous support layer deposited or formed on the column wall. The third type are the adsorption-type open tubular columns, in which a thin layer of the adsorbent is deposited or formed on the inside column wall.

It is further proposed that the liquid-phase deposit should always be referred to as a *film* while any deposited (or formed) solid material should be called a *layer*. Thus, we have a stationary-phase film coated either on the inside column wall or on a thin porous support layer deposited (or formed) on the inside column wall.

Both the support and the adsorbent can either be deposited or formed by chemical treatment of the inside wall of the tubing. Columns prepared by the latter method will be termed *wall-treated* columns.

1.3 TERMS AND DEFINITIONS

In order to follow the discussion in subsequent chapters of this book, an understanding of some elementary concepts is required.

These are discussed at length in various gas chromatography textbooks; here, only their definition and the method of their calculation are given.

1.31 Characterization of a Peak

The chromatographic peak is characterized by its *retention time* (t_R), the time (or distance on the recorder chart) between the point of injection and the peak maximum. The retention time of an inert component (the "air peak time" or the *gas hold-up* of the column) is denoted by t_M. The retention time measured from the maximum of the inert gas peak is called the *adjusted retention time* (t_R'):

$$t_R' = t_R - t_M \tag{1}$$

The peak itself is characterized by its width at different heights, e.g., the *half width* (w_h) (measured at half height) and the base width or *base intercept* (w_b) (the distance along the baseline between the intercepts of the tangents drawn through the inflection points of the Gaussian peak) or the *peak width at the inflection points* (w_i).

1.32 The Partition Process

When a sample is injected into a gas chromatographic system, it is carried through the column by the carrier gas. If an inert component is present in the sample (one which does not dissolve in the stationary phase), it will go through the column with the velocity of the carrier gas; on the other hand, the other sample components will be retarded for longer or shorter times by the partition process. The partition process is defined by the *partition coefficient* (K) equal to the equilibrium ratio of the concentrations of the sample component in equal volumes of the stationary and gas phases, respectively. At the same time, the *partition (capacity) ratio* (k) expresses the equilibrium ratio of the *amounts* of sample component in the stationary and gas phases of the column:

$$k = \frac{t_R - t_M}{t_M} = \frac{t_R'}{t_M} \tag{2a}$$

thus,

$$t_R = t_M(1 + k) \tag{2b}$$

The partition ratio is thus equal to the ratio of the times an average molecule of sample component spends, respectively, in the stationary and gas phases in its passage through the column.

The values of K and k are related by the so-called β *term*, which is the ratio of the total gas volume in the column (V_G) to the total volume of the stationary phase (V_L), at column temperature. The β term characterizes the column itself:

$$\beta = \frac{V_G}{V_L} \tag{3}$$

$$K = \beta k \tag{4}$$

It should be mentioned that in the European literature, the symbol k is often used for the partition coefficient and the symbol k' for the partition ratio. In this book, the symbols K and k will be used consistently.

1.33 Peak Separation

The relative position of two consecutive peaks in a chromatogram is expressed by the *relative retention* or *relative volatility* (α):[10]

$$\alpha = \frac{t'_{R2}}{t'_{R1}} = \frac{K_2}{K_1} = \frac{k_2}{k_1} \tag{5}$$

where, by definition, $t'_{R2} > t'_{R1}$.

The relative volatility does not express the true separation of two consecutive peaks because it does not take into account the sharpness of the two peaks. The true separation of two consecutive peaks is expressed by the *peak resolution*:

$$R = \frac{t'_{R2} - t'_{R1}}{\frac{1}{2}(w_{b1} + w_{b2})} = \frac{2\Delta t}{w_{b1} + w_{b2}} \tag{6}$$

If the two peaks are fairly close to each other, $w_{b1} \simeq w_{b2}$ and

$$R \simeq \frac{\Delta t}{w_{b2}} \tag{7}$$

If $R = 1$, the resolution of two equal-area peaks is about 98% complete, while if $R = 1.5$, the resolution is 99.7% complete ("base-line separation").

---

[10] The symbol α is generally used if the relative position of two *consecutive* peaks is given, while the symbol r is used to refer to the retention relative to *one* particular component (standard).

1.34   Column Efficiency

The efficiency of a column is expressed by the *number of theoretical plates (n)*:

$$n = 16\left(\frac{t_R}{w_b}\right)^2 = 5.54\left(\frac{t_R}{w_h}\right)^2 \tag{8}$$

The theoretical plate is a hypothetical part of the column in which equilibrium is established during the partition process. The length of such a hypothetical unit (the *height equivalent to one theoretical plate,* HETP) can be calculated from the column length (L) and the number of theoretical plates (n):

$$\text{HETP} = \frac{L}{n} \tag{9}$$

In practical gas chromatography, an expression relating k, α, and R is most important, i.e., the ability to calculate the *number of theoretical plates required* (n_{req}) in order to achieve a desired resolution (R) for a given component pair defined by α and k. Such an equation was derived by Purnell [225, 226] for two relatively closely positioned peaks:

$$n_{req} = 16R^2\left(\frac{\alpha}{\alpha - 1}\right)^2\left(\frac{k + 1}{k}\right)^2 \tag{10}$$

In this equation, k refers to the second peak.

By substituting K/β for k, we can write Eq. (10) in the following form:

$$n_{req} = 16R^2\left(\frac{\alpha}{\alpha - 1}\right)^2\left(\frac{\beta}{K} + 1\right)^2 \tag{11}$$

In Eqs. (10) and (11) the number of theoretical plates is calculated. Other possibilities are to calculate the *maximum possible resolution* (R^*) for a given theoretical plate number or the *smallest relative volatility* (α^*) which still permits a certain separation of a component pair:

$$R^* = \frac{\sqrt{n}}{4}\left(\frac{\alpha - 1}{\alpha}\right)\left(\frac{k}{k + 1}\right) \tag{12a}$$

$$\alpha^* = \frac{\sqrt{n}}{\sqrt{n} - 4R\left(\dfrac{k+1}{k}\right)} \tag{12b}$$

Similar equations can also be derived from Eq. (11).

1.35   Flow Rate and Gas Velocity

During gas chromatographic analysis, a carrier gas is flowing through the separation column. In most cases, the *flow rate of the carrier gas* (F_a) is kept constant. If the analysis is carried out under isothermal conditions, the flow rate is kept constant by maintaining a constant pressure drop (Δp) through the column. In programmed-temperature analysis, in order to keep F_a constant, the inlet pressure (i.e., the pressure drop) has to be changed; this is accomplished with the help of automatic flow regulators. A special technique proposed for programmed-temperature conditions is to keep the pressure drop and not the flow rate constant during analysis; also recently, programming of the flow rate instead of the temperature has been proposed. These special techniques are discussed in the fourth part of this book.

The flow rate of the carrier gas (F_a) is usually measured at the column end, at column outlet pressure and ambient temperature. If the flow rate is measured with a bubble flowmeter, a correction should be made for the vapor pressure of water. The proper *corrected flow rate* (F_c), which is corrected to column temperature and true outlet pressure, can be calculated by using Eq. (13):

$$F_c = F_a \frac{T_c}{T_a}\left(1 - \frac{p_w}{p_a}\right) \tag{13}$$

where T_a and T_c are the ambient and column temperatures (in °K), and p_a and p_w are, respectively, the ambient pressure and the partial pressure of water at outlet temperature.

In the theoretical treatment concerning column efficiencies, the *carrier gas velocity* is used rather than the flow rate. The carrier gas velocity at column outlet (u_o) (corrected to column temperature) can be calculated from the flow rate. In the case of open tubular columns, u_o is calculated by dividing F_c by the cross-sectional area of the column tubing:

$$u_o = \frac{F_c}{r^2\pi} \tag{14}$$

where r is the column radius.

In the case of packed columns, the situation is more complicated because here the entire cross-sectional area of the column is not available to the moving gas. The fraction of column cross section available for the moving carrier gas is called the *interparticle porosity* (ε); thus, the outlet gas velocity is calculated using Eq. (15):

$$u_o = \frac{F_c}{\varepsilon r^2\pi} \tag{15}$$

For proper calculation, one more correction has to be made. Since a pressure drop (Δp) exists along the column, the pressure and velocity of the carrier gas will vary along the length of the column. Therefore, the flow rate and linear gas velocity measured at column outlet should be corrected to average conditions in the column. This is done by the application of the *compressibility correction factor* (j) originally derived by James and Martin:

$$j = \frac{3}{2}\frac{(p_i/p_o)^2 - 1}{(p_i/p_o)^3 - 1} \tag{16}$$

where p_i and p_o are the absolute carrier gas inlet and outlet pressures. In practical calculation, the carrier gas outlet pressure (p_o) is usually taken as equal to the ambient pressure (p_a). Figure 2 plots j against p_i/p_o and—assuming that $p_o = p_a = 1$—against Δp.

The *average linear carrier gas velocity* $(\bar{u})$ and the corresponding average flow rate $(\bar{F})$ can thus be calculated from Eqs. (17a) and (17b):

$$\bar{u} = u_o j \tag{17a}$$

$$\bar{F} = F_c j \tag{17b}$$

In some modern commercial gas chromatographs, the inlet pressure is not measured directly. This is the case, for example, where automatic flow regulators are used and a pressure drop exists through the regulator which is not indicated. In such cases, the average linear gas velocity can be calculated from the column length (L) and the retention time of an inert component (t_M), assuming, of course, that

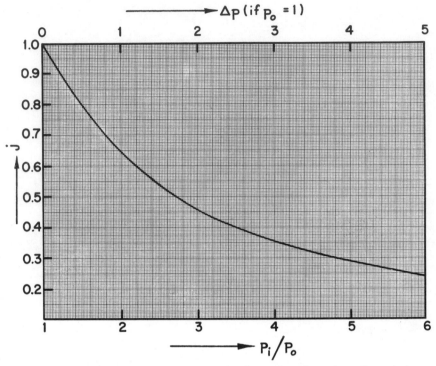

Figure 2. Plot of the compressibility correction factor (j) against p_i/p_o and—assuming that $p_o = 1$ atm—against Δp.

the volume of "dead space" in the instrument is negligible compared to the volume of the column:

$$\bar{u} = \frac{L}{t_M} \tag{17c}$$

Finally, the relationship between the retention time of a particular substance defined by the capacity ratio (k), the column length (L), and the average linear gas velocity ($\bar{u}$) has to be expressed. Substituting t_M from Eq. (17c) into Eq. (2b), we obtain

$$t_M = \frac{L}{\bar{u}}(1 + k) \tag{18}$$

Second Part

Theory and Practice

2.1 THE BASIC GOLAY EQUATION

The theory of coated open tubular columns developed by Golay [109, 110] results in the following equation for HETP in terms of the average linear gas velocity ($\bar{u}$):

$$\text{HETP} = \frac{B}{\bar{u}} + C_G \bar{u} + C_L \bar{u} \tag{19}$$

In this basic equation, B represents the longitudinal gaseous diffusion while C_G and C_L are the resistance to mass transfer in the gas and liquid phases respectively. These three terms can be expressed with the help of the coefficients of gaseous (D_G) and liquid diffusion (D_L), the inside column radius (r), the partition coefficient (K), and the partition ratio (k):

$$B = 2D_G \tag{20}$$

$$C_G = \frac{1 + 6k + 11k^2}{24(1 + k)^2} \frac{r^2}{D_G} \tag{21}$$

$$C_L = \frac{k^3}{6(1 + k)^2} \frac{r^2}{K^2 D_L} \tag{22}$$

Equation (22) can be modified by expressing k in terms of Eq. (4):

$$C_L = \frac{K}{(K + \beta)^2} \frac{r^2}{6\beta D_L} \tag{23}$$

A further modification of Eq. (22) introduces the thickness of the liquid phase (d_f). This is done by calculating the value of β from column geometry:

$$\beta = \frac{V_G}{V_L} = \frac{(r - d_f)^2}{r^2(r - d_f)^2} = \frac{(r - d_f)^2}{d_f(2r - d_f)} \tag{24}$$

However, since $r \gg d_f$, Eq. (24) can be simplified:

$$\beta = \frac{r}{2d_f} \tag{25}$$

Substituting this value into Eq. (4), we obtain

$$K = \frac{r}{2d_f}k \tag{26}$$

Substitution of this value of the partition coefficient into Eq. (22) gives

$$C_L = \frac{2k}{3(1 + k)^2} \cdot \frac{d_f^2}{D_L} \tag{27}$$

A comparison of Eq. (19) with various forms of the so-called van Deemter equation shows that the multiple-path term of the latter is missing since in open tubular columns there is only one gas path. The B term of Eq. (19) is equal to the molecular diffusion term of the van Deemter equation, except that the turtuosity factor does not appear here because for open tubular columns it is equal to unity. The C_L term—as given in Eq. (27)—is equal to the corresponding term of the extended van Deemter equation,[11] while the C_G term has the same form as the corresponding term of the extended van Deemter equation if the column radius is substituted for particle diameter and gas film thickness as in the case with open tubular columns.

2.2   PRACTICAL   CONCLUSIONS   FROM   THE   GOLAY EQUATION

Based on the Golay equation, many detailed studies have been carried out on the influence of the individual parameters on column efficiency. In the following chapters, some of the most important questions will be discussed briefly. Readers who are interested in studying these and other problems associated with the theory of

---

[11] S. Dal Nogare and R. S. Juvet, Jr., *Gas Liquid Chromatography—Theory and Practice.* Interscience, New York, 1962, pp. 95–100.

Golay columns in greater detail are referred to the publications of Desty *et al.* [63–66], Giddings [102–104], Jentzsch and Hövermann [153], Kaiser and Struppe [162], Khan [168], Knox [172, 173], Norem [204], Purnell and Quinn [227], Scott and Hazeldean [130, 247, 249], Struppe [260, 262], and Thijssen [271].

2.21 Column Efficiency

Plotting HETP against $\bar{u}$ from Eq. (19), we obtain the hyperbolic curve of the van Deemter plot. This hyperbola has a minimum at

$$\text{HETP}_{\text{min}} = 2\sqrt{B(C_G + C_L)} \qquad (28)$$

The relative magnitude of C_G and C_L is dependent on the partition coefficient (K) and the column geometry (expressed by β). For low values of β and relatively low values of K, C_L dominates the resistance to mass transfer, while for high values of β, C_G dominates it and C_L can be considered very small compared to C_G. If C_L is negligibly small, Eq. (28) can be written by substituting the proper expressions for B and C_G from Eqs. (20) and (21):

$$\text{HETP}_{\text{min}} = 2\sqrt{BC_G} = r\sqrt{\frac{1 + 6k + 11k^2}{3(1 + k)^2}} = rf \qquad (29)$$

Figure 3 plots f against k; the corresponding numerical values are listed in Table I. The HETP$_{\text{min}}$ value for a peak with $k = 0$

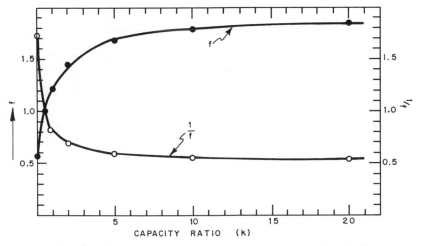

Figure 3. Plots of the values of f and $1/f$ against the capacity ratio (k).

TABLE I
Numerical Values of Some Important Functions of k

k	f	$\dfrac{1}{f}$	$\dfrac{k+1}{k}$	$\left(\dfrac{k+1}{k}\right)^2$	$\dfrac{k}{k+1}$	$\left(\dfrac{k}{k+1}\right)^2$
0	0.577	1.733	—	—	—	—
0.10	0.666	1.501	11.000	121.000	0.091	0.008
0.25	0.825	1.212	5.000	25.000	0.200	0.040
0.50	1.000	1.000	3.000	9.000	0.333	0.111
1.00	1.225	0.816	2.000	4.000	0.500	0.250
2.00	1.453	0.688	1.500	2.250	0.667	0.445
5.00	1.683	0.594	1.200	1.440	0.833	0.694
10.00	1.788	0.559	1.100	1.210	0.909	0.826
20.00	1.849	0.541	1.050	1.102	0.952	0.906
50.00	1.888	0.530	1.020	1.040	0.980	0.961
100.00	1.900	0.529	1.010	1.020	0.990	0.980

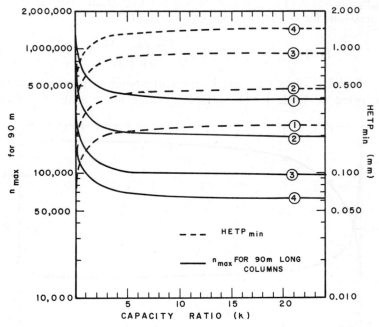

Figure 4. Plots of $HETP_{min}$ for four open tubular columns with different internal diameters and the corresponding n_{max} values for 90-meter-long columns against the capacity ratio (k). Internal diameters: *1* 0.25 mm, *2* 0.50 mm, *3* 1.00 mm, *4* 1.44 mm.

TABLE II

Values of HETP$_{min}$ and n_{max} for Four Different Columns of 90 Meter Length

k	ID = 0.25 mm		ID = 0.50 mm		ID = 1.0 mm		ID = 1.55 mm	
	HETP$_{min}$, mm	n_{max}	HETP$_{min}$, mm	n_{max}	HETP$_{min}$, mm	n_{max}	HETP$_{min}$, mm	n_{max}
0	0.072	1,250,000	0.144	625,000	0.288	312,500	0.447	201,300
0.10	0.083	1,084,400	0.166	542,200	0.333	270,300	0.516	174,400
0.25	0.103	873,800	0.206	436,900	0.412	218,400	0.639	140,800
0.50	0.125	720,000	0.250	360,000	0.500	180,000	0.775	116,100
1.00	0.153	588,200	0.306	294,100	0.612	147,100	0.949	94,800
2.00	0.182	492,500	0.363	247,900	0.726	124,000	1.126	79,900
5.00	0.210	428,600	0.421	213,300	0.841	107,000	1.304	69,000
10.00	0.224	401,800	0.447	201,300	0.894	100,700	1.386	64,900
20.00	0.231	389,600	0.462	194,800	0.924	97,400	1.433	62,800
50.00	0.236	381,400	0.472	190,700	0.944	95,300	1.463	61,500
100.00	0.238	378,100	0.475	189,500	0.950	94,700	1.473	61,100

(the so-called "air peak") is equal to $0.58r$, while for large k values, HETP_{min} approaches $1.9r$.

Equation (29) shows that HETP_{min} is directly proportional to the column radius. This means that columns with larger diameter (assuming the same k value) will show lower efficiency. Figure 4 plots the HETP_{min} values for four columns with various internal diameters and the corresponding maximum achievable theoretical plate numbers (n_{max}) for 90-meter-long columns; the corresponding numerical values are listed in Table II.

HETP_{min} and n_{max} are theoretical values which cannot be reached in practice, mainly for four reasons:

(a) In writing Eq. (29) the C_L term is neglected although it is not equal to zero;

(b) Ideal columns have been assumed, with a completely uniform film thickness and constant diameter along the entire length;

(c) It has been assumed that ideal gas chromatographic systems are used in which the width of the peak is a function of only the partition in the column;

TABLE III
Utilization of Theoretical Column Efficiency with Three Different Columns[a] [83]

ID of column, mm	k	HETP_{min}, mm	HETP, mm	Utilized fraction of theoretically best performance, %
0.25	0.5[b]	0.12	0.20	60
	1.0[b]	0.15	0.27	55
	5.0	0.21	0.66	32
	10.0	0.22	0.74	30
0.50	0.5	0.25	0.70	36
	1.0	0.31	1.30	24
	4.0	0.41	1.93	21
	5.0	0.42	1.99	21
1.55	0.5	0.8	2.0	40
	1.0	0.9	4.2	21
	4.0	1.3	7.3	18

[a] The corresponding chromatograms and analysis conditions are given in Figures 5–7.
[b] Extrapolated values.

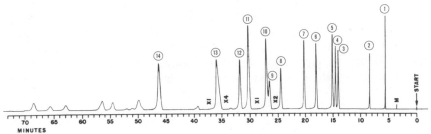

Figure 5. Analysis of benzene homologs from a natural mixture [85]. Column: 150 ft × 0.25 mm ID open tubular, coated with meta-bis(*m*-phenoxyphenoxy)benzene + squalane (8/2, w/w) mixed liquid phase. Column temperature: 65°C. Carrier gas (N₂) inlet pressure, flow rate at outlet, and average linear gas velocity: 20 psi (gauge), 1.5 ml/min, 29.5 cm/sec. Sample size: 1 µl solution, split 1/250. Identified peaks: *M* methane, *1* benzene, *2* toluene, *3* ethylbenzene, *4* *p*-xylene, *5* *m*-xylene, *6* *o*-xylene, *7* isopropylbenzene, *8* *n*-propylbenzene, *9* *p*-ethyltoluene, *10* *m*-ethyltoluene, *11* 1,3,5-trimethylbenzene, *12* *o*-ethyltoluene, *13* 1,2,4-trimethylbenzene, *14* 1,2,3-trimethylbenzene. Peaks after *14* include C₁₀–C₁₄ aromatics, among them isobutylbenzene, 1-methyl-3-isopropylbenzene, 1-methyl-4-isopropylbenzene, 1,3-dimethyl-5-ethylbenzene, and indane.

(d) It has been assumed that one is working at optimum gas velocities, although during practical analysis this is almost never the case.

Experience shows that in practice only 20–60% of the theoretically obtainable maximum plate number is utilized. The utilization is generally better for earlier peaks and for columns with smaller diameters than for peaks with large *k* and/or for open tubular columns with larger diameters. Measurements carried out in the author's

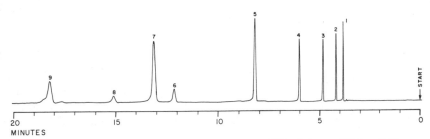

Figure 6. Analysis of a fatty acid methyl ester mixture [82]. Column: 300 ft × 0.50 mm ID open tubular, coated with butanediol succinate liquid phase. Column temperature: 196°C. Carrier gas (He) inlet pressure, flow rate at outlet, and average linear gas velocity: 22.5 psi (gauge), 10 ml/min, 45.8 cm/sec. Sample size: 1 µl (liquid), split 1/60. Peaks: methyl *1* caprylate, *2* caprate, *3* laurate, *4* myristate, *5* palmitate, *6* stearate, *7* oleate, *8* linoleate, *9* linolenate.

laboratory [83], summarized in Table III, showed that for a 0.25 mm ID column the utilization was 50–60% for peaks with $k < 1$ and 30% for peaks with $k > 5$. For columns having an internal diameter of 0.50 mm or larger, the range of utilization was 40–20%. The corresponding three chromatograms are shown in Figures 5–7.

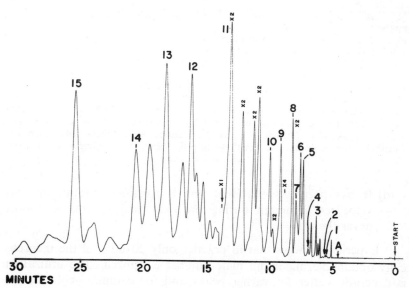

Figure 7. Analysis of a "light virgin naphtha" sample [83]. Column: 300 ft × 1.55 mm ID open tubular, coated with squalane liquid phase. Column temperature: 60°C. Carrier gas (He) flow rate at outlet and average linear gas velocity: 30.6 ml/min, 26.2 cm/sec. Sample size: 1 μl (liquid), no split. Identified peaks: *A* air, *1* isopentane, *2* n-pentane, *3* n-hexane, *4* benzene, *5* cyclohexane, *6* 2,3-dimethylpentane, *7* 2,2,4-trimethylpentane, *8* n-heptane, *9* methylcyclohexane, *10* toluene, *11* n-octane, *12* ethylbenzene, *13* m/p-xylene, *14* o-xylene, *15* n-nonane.

2.22   Carrier Gas Flow

From the Golay equation the optimum average linear gas velocity ($\bar{u}_{opt}$) corresponding to HETP$_{min}$ may be expressed as

$$\bar{u}_{opt} = \sqrt{\frac{B}{C_G + C_L}} \tag{30}$$

Again, if C_L is assumed to be negligibly small, $\bar{u}_{opt}$ can be calculated with the help of the following equation:

$$\bar{u}_{\text{opt}} = \sqrt{\frac{B}{C_G}} = 4\frac{D_G}{r}\sqrt{\frac{3(1+k)^2}{1+6k+11k^2}} = 4\frac{D_G}{r}\frac{1}{f} \qquad (31)$$

The two limiting values of $\bar{u}_{\text{opt}}$ for $k = 0$ and $k = \infty$ are $6.9 D_G/r$ and $2.1 D_G/r$, respectively.

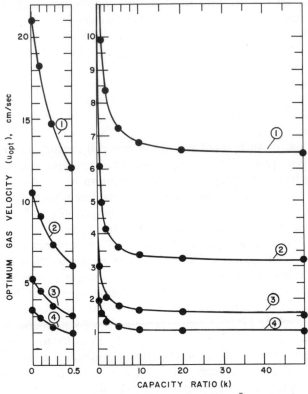

Figure 8. Plot of $\bar{u}_{\text{opt}}$ and the corresponding $\bar{F}_{\text{opt}}$ values for n-heptane samples on four open tubular columns with different internal diameters against the capacity ratio (k), using nitrogen as carrier gas. Internal diameters: *1* 0.25 mm, *2* 0.50 mm, *3* 1.00 mm, *4* 1.55 mm.

Scott and Hazeldean [249] reported the following D_G values for n-heptane as a typical sample, in different carrier gases:

hydrogen: $0.095 \text{ cm}^2/\text{sec}$
nitrogen: $0.038 \text{ cm}^2/\text{sec}$
argon: $0.039 \text{ cm}^2/\text{sec}$

It is evident that the values of $\bar{u}_{opt}$ calculated from Eq. (31) will be small. As an example, Figure 8 plots the $\bar{u}_{opt}$ (and the corresponding optimum average flow rate) values for open tubular columns with four different radii, for n-heptane, using nitrogen as carrier gas. As Desty [65] later pointed out, the values of D_G are actually dependent on the pressure drop across the column, although with low pressure drops the variation is relatively insignificant. Thus, even if the values plotted in Figure 8 are not exact, they are close enough to the actual values so that they can illustrate the optimum velocity ranges.

As seen in Eq. (31) (and Figure 8), the optimum gas velocity is a function of k. This means that if it is adjusted for early peaks, then peaks with larger k values will be far off the optimum, and *vice versa.*

The use of such small gas flows naturally results in much too long analysis times. Therefore, in practical analysis, one is usually operating with flow rates higher than optimum.

It is interesting to note that with velocities slightly higher than optimum, an increase in the carrier gas velocity with a simultaneous increase in column length will reduce the time of analysis while still maintaining the overall efficiency of the column. Scott and Hazeldean [249] gave the following example for a $100 \, \text{ft} \times 0.50 \, \text{mm}$ ID open tubular column operated with hydrogen carrier gas where $\bar{u}_{opt}$ was found to be 16 cm/sec for the n-heptane peak. If the average linear gas velocity was doubled, thus halving the elution time [see Eq. (18)], the HETP was only increased by 25%; therefore, if the column length was also increased by 25% in order to maintain the same efficiency and resolution, the elution time for a given substance was still reduced by a factor of $\frac{1}{2} \times \frac{5}{4} = \frac{5}{8}$. This gas velocity range corresponds to the curved part of the HETP *vs.* $\bar{u}$ plot (van Deemter plot). According to Scott and Hazeldean, the limiting value is that point where the van Deemter plot starts to show a linear relationship between HETP and $\bar{u}$. The transition point between the curved and the linear parts of the plot is the *optimum practical gas velocity* (OPGV), where by proper selection of column length the same overall efficiency can be obtained—but in a shorter time—than with a shorter column operated at $\bar{u}_{opt}$.

At the OPGV, the B term (the longitudinal gaseous diffusion) becomes very small, and thus the first term on the right-hand side of the Golay equation can be neglected. This means that now the

HETP is primarily governed by the resistance-to-mass-transfer terms:

$$\text{HETP} \simeq (C_G + C_L)\bar{u} \tag{32}$$

On the other hand, as given by Eqs. (21) and (22), both resistance-to-mass-transfer terms are directly proportional to the *square* of column radius. This means that when working at the OPGV, a *reduction* of the column radius would actually be more efficient than an increase in column length as shown above. However, since reduction of column radius would very soon necessitate excessive inlet pressures, practical and economic considerations favor the increase of column length for maintaining the same efficiency.

The OPGV concept is very useful in practical analysis. However, it should be regarded primarily as an operating parameter and not as a design parameter.

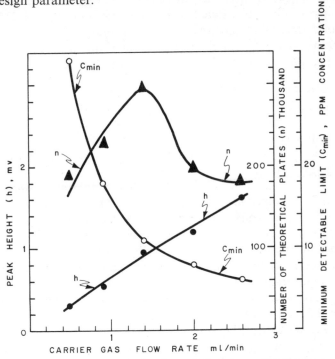

Figure 9. Relationship of the theoretical plate number (n), peak height (h), and the minimum detectable limit (concentration, c_{min}) for a 150 ft × 0.25 mm ID open tubular column with Ucon Oil 550–X liquid phase, calculated for the peak of cyclopentane.

If the average gas velocity is increased above the OPGV, the analysis time will be decreased even more, but now at the expense of a decrease in column efficiency that can no longer be corrected by an increase in length. However, since—as will be seen later—open tubular columns usually have a much higher intrinsic efficacy than actually needed, it is not unusual to operate these columns with gas velocities significantly above the OPGV.

A second reason for operating at higher gas velocities (flow rates) concerns the detectors: if a mass-sensitive detector (such as the flame ionization detector) is used, higher flow rates result in taller peaks, thus increasing the minimum detectable limit of the system. Figure 9 illustrates this basic relationship by plotting the respective values from a series of investigations [81]. A 150 ft × 0.25 mm ID column coated with Ucon Oil 550-X liquid phase was used and operated at 100°C in conjunction with a flame ionization detector. A sample of 1 μl was injected with a split ratio of about 1/250. The minimum detectable limit refers to cyclopentane and is expressed as a concentration (c_{min}) in the sample equivalent to a peak height of twice the noise level. In this case, the noise level was 5 microvolts.

2.23   Selection of Carrier Gas

Among the terms in the Golay equation, both B (the longitudinal gaseous diffusion term) and C_G (the resistance to mass transfer in the gas phase) contain the gaseous diffusion coefficient (D_G). However, this value has an opposite influence on the two basic terms: it is directly proportional to B but inversely proportional to C_G.

It has been seen in the preceding chapter that above the OPGV the B term becomes negligible and the HETP is primarily governed by the C_G term, while below the OPGV the influence of B on the HETP can be significant. This means that the selection of carrier gas is principally dependent on the gas velocities used.

In the preceding chapter, some values of D_G were listed for a typical hydrocarbon sample. Generally, it can be stated that a carrier gas with low density (e.g., hydrogen or helium) allows a higher diffusivity (i.e., larger D_G values) for the sample molecules, while the diffusivity (the value of D_G) in gases with higher density (e.g., nitrogen or argon) is low. Thus, the following general rules for carrier gas selection can be stated:

(a) When working at or above the OPGV, where the C_G term is dominant and the B term is negligible, one should use a carrier gas of low density in order to increase the diffusivity of the solute (the sample molecules) in the gas phase. In this way, D_G will become larger and, consequently, the C_G term will be small.

(b) When working at gas velocities below the OPGV, the B term will soon dominate. Therefore in such cases, it is preferable to use a carrier gas of higher density. In this way, D_G will be low and, thus, the value of the B term can be kept small.

Another case where the use of a higher-density carrier gas is beneficial is when C_L is limiting, because in this case it permits reduction of this term without excessive static diffusion.

However, one should not forget that if a carrier gas of higher density (low D_G) is used, then—as a consequence of Eq. (31)—the value of $\bar{u}_{opt}$ will also decrease. This means that if one wants to achieve the smallest possible HETP, the analysis time will become even longer. This fact again militates against working at theoretical optimum conditions in practical analysis.

Since with open tubular columns—as discussed in chapter 2.22—one is usually operating with velocities at or above the OPGV, the use of hydrogen or helium as carrier gas is definitely preferred. When using nitrogen, a somewhat poorer performance will be obtained. However, since the overall efficiency of the open tubular columns is generally very high, this reduction of performance usually does not restrict their application.

2.24   Liquid-Film Thickness

The influence and proper choice of the liquid-film thickness (d_f) are more complicated than of any other column parameter. The reason for this is that here practical considerations are even more important than, e.g., in selecting the proper flow rates.

It has been pointed out previously that the resistance to mass transfer in the liquid-phase term (C_L) should be kept as low as possible. With both the internal diameter and the partition coefficient constant, it follows from Eqs. (25) and (4) that reduction of d_f will increase the value of β and thus decrease the value of the partition ratio (k). On the other hand, as seen from Eq. (27), decrease of d_f and k will result in a decrease of the value of C_L. This means that one should favor a very small liquid-film thickness. As shown by Desty and Goldup [63], the reduction of d_f has the greatest influence on the overall C term $(C_G + C_L)$ in the case of samples of low molecular weight.

Another reason why a smaller liquid-film thickness is desired is connected with the time of analysis. It is evident that with thicker films, the partition process will take longer, thus lengthening the analysis time.

These two reasons favor the smallest possible liquid-film thickness. On the other hand, two further reasons tell us to keep the value of d_f at a reasonable level.

The first reason concerns the values of k. As mentioned above, a reduction in d_f results in large β values. However, open tubular columns already have β values larger than packed columns (see chapter 2.31). If the value of β becomes larger, then—for the same sample component—the partition ratio will naturally become smaller. But it will be shown in chapter 2.32 that lower values of k mean that more theoretical plates are necessary to obtain the same resolution of two consecutive peaks.

The second reason for reasonable liquid-film thicknesses has to do with the sample capacity of the column. Open tubular columns have a generally very low sample capacity, which makes sample introduction rather difficult. Further reduction of the thickness (and therefore the volume) of the liquid phase would even further reduce the sample capacity of the column.

Generally, the liquid-film thickness of open tubular columns used in practical analysis varies between 0.5 and 2.5 μ, and according to general experience the best practical results have been obtained with films about 0.6–1.5 μ thick. Special columns with very high efficiency were fabricated with film thicknesses of about 0.2–0.4 μ. Liquid films thicker than about 2.5 μ are no longer stable.

2.3 COMPARISON WITH PACKED COLUMNS

Let us now investigate the fundamental differences between packed and open tubular columns. The most characteristic differences can be observed in the values of the β term and the permeability of the columns. The β term influences the number of theoretical plates necessary for a given separation, while the permeability controls the time of analysis and sets the practical limits on column length. Thus, in comparing open tubular columns with packed columns, the following questions have to be investigated:

(a) The changes in the β values and their influence on column characteristics;

(b) Variations in the number of theoretical plates necessary for a given separation;

(c) Permeability of the two column types and practical limits due to excessive inlet pressures necessary for maintaining the desired flow.

Subsequently, the performance index and various modified expressions related to the plate number concept will be discussed. These terms have been developed in order to allow a direct comparison of the performance of packed and open tubular columns.

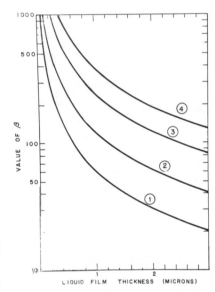

Figure 10. Plots of the values of β against the liquid-film thickness for four open tubular columns with different radii: *1* 125 μ, *2* 250 μ, *3* 500 μ, *4* 775 μ.

2.31 Influence of the β Values on Column Characteristics

As given in Eq. (3), the β term is the ratio of the total gas volume in the column (V_G) to the total volume of the stationary phase (V_L) at column temperature. In packed columns the value of β is relatively small, the reported values[12] for practical columns being in the range from about 6 to 35. On the other hand, in open tubular columns the β values are larger. Figure 10 plots the values of β [calculated from Eq. (25)] for open tubular columns with four different radii (corresponding to 0.25, 0.50, 1.00, and 1.55 mm ID, respectively) against

---

[12] S. Dal Nogare and J. Chiu, *Anal. Chem.* **34**:890 (1962).

TABLE IV
The Values of β for Various Liquid-Film Thicknesses and Column Radii

Liquid-film thickness, μ	Column radius, μ			
	125	250	500	775
	Values of β			
0.10	625	1250	2500	3875
0.25	250	500	1000	1550
0.50	125	250	500	775
0.75	83.3	167	333.3	516.7
1.00	62.5	125	250.0	387.5
1.25	50.0	100	200.0	310.0
1.50	41.7	83.3	166.7	258.3
1.75	35.7	71.4	142.9	221.4
2.00	31.25	62.5	125.0	193.8
2.25	27.8	55.6	111.1	172.2
2.50	25.0	50.0	100.0	155.0
2.75	22.7	45.5	91.0	140.9
3.00	20.8	41.7	83.3	129.2

film thickness; the corresponding numerical values are listed in Table IV. Since in practice the film thickness is generally in the range of 0.25–1.5 μ, the practical β values vary between about 50 and 1500.

At a given temperature, the partition coefficient (K) is independent of the column used; thus, if β becomes larger (i.e., when an open tubular column is used), k must become smaller in order to fulfill Eq. (4). On the other hand, the adjusted retention time (t'_R) is governed solely by the liquid phase; thus, if k becomes smaller, the "air time" (t_M) will simultaneously become relatively larger.

Due to the fact that the partition coefficient is independent of the type of column used when working at identical temperatures (with, of course, the same liquid phase), the value of α will also be the same for both packed and open tubular columns prepared with the same liquid phase and used at the same temperature.

2.32   Changes in the Necessary Column Efficiency

2.321   Number of Theoretical Plates Required at Various α and k Values. Equation (10) expresses the number of theoretical plates required for a desired resolution (R) of two peaks characterized

by α and k. It is evident that if k becomes smaller while R and α
remain the same, n_{req} will be larger. In other words, more theoretical
plates are required on the open tubular column for the same separation as compared with the packed column. This is also evident from
Eq. (11): since β is larger for the open tubular column while K and
α remain the same, the last term on the right-hand side will also
become larger.

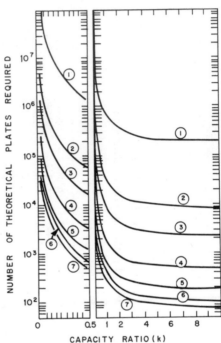

Figure 11. Plots of n_{req} against the capacity ratio (k), at $R = 1.0$, for the following values of α: *1* 1.01, *2* 1.05, *3* 1.10, *4* 1.25, *5* 1.50, *6* 1.75, *7* 2.00. The left side of the diagram is an enlargement of the region up to $k = 0.5$.

Figure 11 plots the n_{req} values for different relative retentions
(α) as a function of k for a resolution of $R = 1$; the corresponding
numerical values are listed in Table V. As an example, let us investigate the situation at $\alpha = 1.10$. If, e.g., the k for the packed column
were 10 and for the open tubular column—due to change in the
β value—0.1, then one could achieve the same resolution on a packed
column with 2350 plates as on an open tubular column with 234,250
plates.

In the above example, the ratio of the k values was taken as 100.
But such a high ratio is very unlikely and—as was seen earlier—

TABLE V
Values of n_{req} Calculated from Eq. (10) for Different Values of α and k

k	$\alpha = 1.01$	$\alpha = 1.05$	$\alpha = 1.10$	$\alpha = 1.25$	$\alpha = 1.50$	$\alpha = 1.75$	$\alpha = 2.00$	$\alpha = 3.00$	$\alpha = 5.00$
0.05	71,978,260	3,111,696	853,776	176,400	63,500	38,416	28,224	15,876	11,025
0.10	19,749,140	853,776	234,256	48,400	17,424	10,540	7,744	4,356	3,025
0.25	4,080,400	176,400	48,400	10,000	3,600	2,178	1,600	900	625
0.50	1,468,944	63,504	17,424	3,600	1,296	784	576	324	225
0.75	888,620	38,416	10,540	2,178	784	474	348	196	136
1.00	652,864	28,224	7,744	1,600	576	348	256	144	100
1.50	453,378	19,600	5,378	1,111	400	242	178	100	69
2.00	367,236	15,876	4,356	900	324	196	144	81	56
2.50	319,903	13,830	3,795	784	282	170	125	70	49
5.00	235,031	10,161	2,788	576	207	125	92	52	36
7.50	209,642	9,063	2,487	514	184	111	82	46	32
10.00	197,491	8,538	2,343	484	176	105	77	44	30

with practical columns (except columns of relatively large diameter) the ratio is much less. At the same time, the total number of theoretical plates which can be obtained on an open tubular column is generally much higher than the total plate number of a packed column (this question will be discussed in the next chapter). Therefore, even the higher n_{req} value can be achieved relatively easily: actually, the number of theoretical plates *available* is usually higher than actually needed. Thus, separations impossible with packed columns can often be achieved with open tubular columns.

2.322 Comparison of the Two Column Types. Let us assume an open tubular and a packed column with the same liquid phase, being used at the same temperature. The number of theoretical plates required for the separation of a given component pair in either column can be calculated from Eq. (10). If we write this equation for both columns and divide one equation by the other, the $16[\alpha/(\alpha - 1)]^2$ term will cancel since, at identical temperature, the relative retention will be the same on both columns. The resulting equation is

$$\frac{n_T}{n_P} = \left(\frac{R_T}{R_P}\right)^2 \left(\frac{k_P}{k_T}\frac{k_T + 1}{k_P + 1}\right)^2 \tag{33a}$$

where subscripts T and P refer to the open tubular and packed columns, respectively. We can simplify this equation:

$$n_T = n_P F_1^2 F_2^2 \tag{33b}$$

where

$$F_1 = \frac{R_T}{R_P} \tag{34a}$$

$$F_2 = \frac{k_P}{k_T}\frac{k_T + 1}{k_P + 1} \tag{34b}$$

Equation (34a) gives a factor relating the resolutions achieved on the two columns. If they are identical, then $F_1 = 1$ and

$$n_T = n_P F_2^2 \tag{35}$$

The factor F_2^2 specifies how many times more theoretical plates are necessary on an open tubular column than on a packed column in order to achieve the same resolution. The factor actually incorporates the ratio of the β values, since from Eq. (4), for work at the same temperature,

$$\beta_T k_T = \beta_P k_P \tag{36}$$

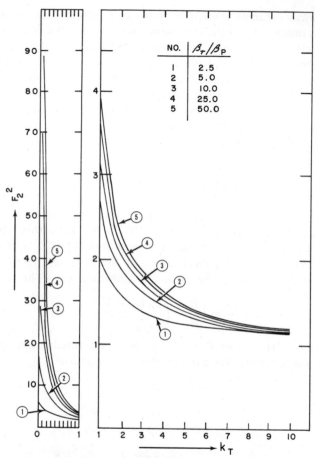

Figure 12. Plots of F_2^2 against the capacity ratio (k_T) on open tubular columns for various β_T/β_P (i.e., k_P/k_T) values. The corresponding k_P values can be obtained by multiplying k_T by the proper β_T/β_P value. The left side of the diagram is an enlargement of the region up to $k_T = 1.0$.

and therefore

$$F_2 = \frac{\beta_T\,k_T + 1}{\beta_P\,k_P + 1} \tag{37}$$

Figure 12 plots the F_2^2 values against k_T for various β_T/β_P (i.e., k_P/k_T) values. It is seen that the multiplication factor is significant only when $k_T < 0.5$ and even then only if the β_T/β_P ratio is large.

The most frequently used open tubular columns have an internal diameter of 0.25 mm while the thickness of the liquid phase is usually between 0.5 and 1.5 μ. This gives a range of 125–42 for β_T (see Figure 10 and Table IV), and if we assume $\beta_P = 15$—which is a fair value—this results in β_T/β_P ratios smaller than 10. Such values are insignificant since—as seen in the next chapter—open tubular columns usually have an intrinsic efficacy much higher than ten times the average efficiency of packed columns. Therefore, F_1 can be greater than unity, i.e., $R_T > R_P$: the higher efficiency can be used for obtaining better resolution with the open tubular columns than is possible with packed columns.

In the case of open tubular columns with internal diameters larger than 0.25 mm, the ratio of the β values will also be larger. Thus, the multiplication factor becomes larger. However—as seen in the next chapter—these columns have such a small pressure drop that columns of relatively great length, with large numbers of theoretical plates, can be used.

2.323 Examples. The following two examples will be used to demonstrate the comparatively much higher overall efficacy of open tubular columns. The first example, the separation of *meta-* and *para*-xylenes, shows that in some cases no separation can be obtained at all on a practical packed column prepared with the same liquid phase, while the second example—the analysis of fatty acid methyl esters—demonstrates that the open tubular column allows a much greater efficiency than a packed column with the same liquid phase.

2.323.1 *Analysis of p/m-Xylenes.* The separation of this component pair was long considered one of the most difficult separation problems in gas chromatography. With packed columns, only a few specific liquid phases could be used for the separation of these isomers. But already very early in the development of open tubular columns was it shown that they can solve this problem much more easily, even if coated with a nonspecific phase. In 1959 Condon [54] used a 175 ft × 0.25 mm ID column coated with poly(propylene glycol) for this purpose, while in 1960 Wiseman [280] showed an almost complete separation of the two isomers on a 50 ft × 0.25 mm ID column prepared with dinonyl phthalate liquid phase. No practical packed column prepared with either phase would separate these two isomers.

A third author who demonstrated the successful application of a standard nonspecific stationary phase for the solution of this problem is Schreyer [239]. He worked at 80°C, with a 140 meter × 0.30 mm ID open tubular column coated with squalane. Since the pertinent numerical values are either listed by him or can readily be measured from the chromatogram, this particular separation can be used as a good example for the comparison of column efficacies.

Table VI lists the numerical values for the open tubular column. The partition coefficients were calculated using Eq. (4). As is seen, 241,000 plates were required to achieve a practically complete separation ($R = 1.37$) of the two peaks.

In order to be able to compare the result obtained on the squalane open tubular column with the performance of a hypothetical squalane packed column, the β value of the latter is needed. According to Dal Nogare and Chiu,[12] the optimum β value for a packed column with 80/100 mesh diatomaceous earth support is 19, with the corresponding liquid phase loading of 8% by weight. This number will be used for the further calculation.

TABLE VI
Comparison of Two Squalane Columns

Parameter	Symbol	Dimension	Open tubular column [239]	Packed column (calculated)
Column length	L	m	140	
Column radius	r	mm	0.15	
Liquid-film thickness	d_f	μ	0.33	
Ratio of gas and liquid volumes in the column	β		454.5	19
Temperature of analysis	T_c	°C	80	80
Retention time of inert gas	t_M	sec	408	
Retention time of p-xylene	t_{R1}	sec	1388	
Retention time of m-xylene	t_{R2}	sec	1404	
Base intercept of the m-xylene peak	w_{b2}	sec	11.71	
Capacity ratio of p-xylene	k_1		2.40	57.41
Capacity ratio of m-xylene	k_2		2.44	58.37
Relative retention	α		1.016	1.016
Partition coefficient of p-xylene	K_1		1090.8	1090.8
Partition coefficient of m-xylene	K_2		1109.0	1109.0
Resolution of p/m-xylene peaks	R		1.37	1.37
Number of theoretical plates required			241,400	126,100

From Eq. (36) the capacity ratio of m-xylene on the hypothetical packed column can be expressed as

$$k_P = \frac{\beta_T}{\beta_P} k_T \tag{38}$$

where the subscripts T and P again refer to the open tubular and packed columns. Since β_T and k_T are known (see Table VI) and β_P was taken as 19, k_P can be calculated. On the other hand, once k_P is known, the number of theoretical plates required to achieve the same separation ($R = 1.37$) of p/m-xylene on the hypothetical packed column can be calculated since the α value will remain the same ($\alpha = 1.016$); the result is 126,100 theoretical plates.

In chapter 2.33, the reality of the two columns will be investigated; it will be seen that no practical packed column with such a performance can be prepared, while with the open tubular column 241,400 plates can be achieved in a relatively short time.

It should be mentioned that Schreyer's chromatogram shows remarkably high efficiency compared to the theoretical best value. For $k = 2.44$ (m-xylene) and for a column of 0.15 mm radius, the value of HETP_{min}—calculated from Eq. (29)—is 0.393 mm, while actually an HETP of 0.580 mm was obtained for the *meta*-xylene peak; this means a yield of 67 %.

2.323.2   Analysis of Fatty Acid Methyl Esters. The previous example showed a case where no practical packed column could result in the same resolution as the corresponding open tubular column. The following example demonstrates that even if a packed column permits the complete separation of two main peaks, one can specify an open tubular column which will yield much greater resolution, allowing the detection of peaks overlapped on the packed column, in the same time.

In the analysis of fatty acid methyl esters, the effectiveness of a column is usually indicated by the separation of the stearate–oleate pair. Figure 13 compares two actual chromatograms showing the separation of these two substances on columns with polyester liquid phase. In these two chromatograms, the packed column was deliberately favored by operating it 10°C lower than the open tubular column; in this way, the α value for the packed column became larger. Table VII lists the corresponding numerical values; as is seen, the number of theoretical plates required to obtain the particular

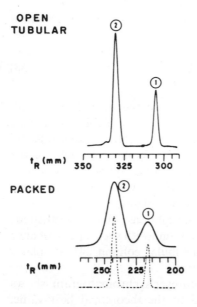

OPEN TUBULAR

t_R (mm) 350   325   300

PACKED

t_R (mm) 250   225   200

Figure 13. Chromatograms of a mixture of *1* methyl stearate and *2* methyl oleate on two columns both prepared with butanediol succinate (BDS) liquid phase. Open tubular column: 150 ft × 0.25 mm ID; packed column: 6 ft × 2.12 mm ID, 8% by weight BDS on silanized Chromosorb W 80/100 mesh. Column temperatures: 180 and 170°C, respectively. For characteristic data, see Table VII.

separation was 3050 on the packed and 47,300 on the open tubular column, with resolution values of 1.2 and 4.1, respectively. This is an illustration of the fact that since the "intrinsic efficiency" of the open tubular column is significantly greater than necessary for obtaining just the same resolution, a much better separation can be obtained.

TABLE VII
Numerical Values Corresponding to the Chromatograms in Figure 13

Parameter	Symbol	Dimension	Open tubular column	Packed column
Retention time of inert gas (methane)	t_M	mm	35.0	30.0
Retention time of stearate	t_{R1}	mm	305.5	218.0
Retention time of oleate	t_{R2}	mm	330.5	239.0
Base intercept of the oleate peak	w_{b2}	mm	6.5	17.2
Capacity ratio of oleate	k		8.4	7.0
Relative retention	α		1.09	1.11
Resolution	R		4.1	1.2
Number of theoretical plates required for the given resolution	n_{req}		47,300	3050

If one wishes to obtain the same resolution on the packed column as on the open tubular column (dotted chromatogram), 35,000 theoretical plates are required. On the other hand, if a resolution of 1.2 were also satisfactory on the open tubular column, a peak having a relative retention as low as $\alpha = 1.026$ could still be separated from the oleate peak [see Eq. (13)].

The increase in resolution or the decrease in the relative retention of peaks which still can be separated is very important in the analysis of practical samples. Packed columns give good separation of the stearate–oleate pair; however, if a large number of isomers are present in the same area of the chromatogram, they will be overlapped by the two main peaks since a practical packed column generally cannot sufficiently separate peaks with a relative retention smaller

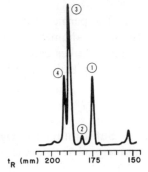

Figure 14. Part of the chromatogram of menhaden oil methyl ester obtained on a 150 ft × 0.25 mm ID open tubular column with BDS liquid phase, at 185°C. Peaks: *1* stearate, *2* ?, *3* oleate, *4* ?.

than 1.10. This is illustrated in Figure 14, which shows a part of the chromatogram in the C_{18} range, obtained from the analysis of menhaden oil methyl ester.[13] Menhaden oil consists of a large number of unsaturated fatty acids. On a packed column the unsaturated C_{18} isomers give a *single* peak, which is well separated from stearate and is taken as methyl oleate; on the other hand, the use of an open tubular column reveals at least three isomers.

2.33 Influence of Column Permeability

2.331 Specific Permeability. A column through which a gas is flowing offers a certain resistance to the gas flow. A convenient parameter for defining this resistance is the specific permeability (B_0)

---

[13] F. J. Kabot (Perkin–Elmer Corp., Norwalk, Conn.), unpublished data.

of the column; for *packed columns*, it is expressed by the Kozeny–Carman equation:[14,15]

$$B_0 = \left(\frac{d_p^2}{180}\right)\left(\frac{\varepsilon^3}{(1 - \varepsilon)^2}\right) = 2\eta\varepsilon L \frac{p_o}{p_i^2 - p_o^2} u_o \tag{39}$$

where d_p is the effective particle diameter, ε the interparticle porosity as discussed in chapter 1.35, η the viscosity of the carrier gas, L the column length, p_i and p_o the inlet and outlet pressures of the carrier gas, and u_o the linear gas velocity at column outlet.

According to measurements reported by Dal Nogare[15] and Bohemen and Purnell,[16] the interparticle porosity of packed columns is remarkably constant and $\varepsilon \simeq 0.40$. Thus,

$$B_0 \simeq \frac{d_p^2}{1012} \tag{40}$$

The permeability of *open tubular columns* can be calculated from the inside diameter (d) of the tubes. The outlet gas velocity of long capillary tubes can be expressed by the Hagen–Poisseuille equation written by Kaiser and Struppe [160] in the following form:

$$u_o = \frac{d^2}{64\eta L} \frac{p_i^2 - p_o^2}{p_o} \tag{41}$$

Substituting the right side of this equation into Eq. (39) and taking $\varepsilon = 1$ (because in open tubular columns, the whole cross section is available to the moving carrier gas), we obtain the following equation for the permeability of open tubular columns with smooth inner surfaces:

$$B_0 = \frac{d^2}{32} = \frac{r^2}{8} \tag{42}$$

where d and r are the inside diameter and radius of the column, respectively. The concept of "specific permeability" in the case of open tubular columns has validity only when a hypothetical bundle of identical capillaries is considered; however, it is convenient to use the same symbol (B_0) in both cases.

---

[14] P. C. Carman, *The Flow of Gases through Porous Media*, Butterworths, London, 1956.

[15] S. Dal Nogare and R. S. Juvet Jr., *Gas Chromatography—Theory and Practice*, Interscience, New York, 1962, pp. 133–136.

[16] J. Bohemen and J. H. Purnell, *J. Chem. Soc.* **1961**:360.

TABLE VIII
Specific Permeability of Various Columns

Type of column	Support	Mesh range		Column ID, mm	Specific permeability, $\times 10^{-7}$ cm²	Source
		Mesh	Particle diameter, μ			
Packed	Silanized Chromosorb R	80/100	177–149	—	1.96	a
		60/80	250–177	—	3.45	
		40/45	420–350	—	9.54	
	Sterchamol	80/100	177–150	—	2.7	b
	Glass beads		177–150	—	4.95	[141]
Open tubular	—	—	—	0.10	31	c
	—	—	—	0.25	195	c
	—	—	—	0.50	781	c
	—	—	—	1.00	3125	c
	—	—	—	1.55	7508	c

[a] S. Dal Nogare and R. S. Juvet, *Gas–Liquid Chromatography—Theory and Practice*, Interscience, New York, 1962, p. 135.
[b] A. I. M. Keulemans, *Gas Chromatography*, 2nd ed., Reinhold, New York, 1959, p. 147.
[c] Calculated from Eq. (42).

The permeability is a characteristic of the column and is independent of the liquid phase used; in packed columns it is proportional to the effective particle diameter,[17] and in open tubular columns to the inside diameter of the tubing. Table VIII lists permeability values of various columns.

2.332 Correlation Between Specific Permeability and Pressure Drop Through the Column.

The pressure drop (Δp) through the column necessary to maintain a given average linear carrier gas velocity ($\bar{u}$) is expressed by the following equation:

$$\Delta p = \frac{\eta}{B_0} L\bar{u} \tag{43}$$

where η is the viscosity of the carrier gas at column temperature, B_0 is the specific permeability discussed in the previous chapter, and L is the column length. If the viscosity of the carrier gas is expressed in poises (gm $\cdot$ cm^{-1} $\cdot$ sec^{-1}), the length of the column in cm, and the average linear gas velocity in cm $\cdot$ sec^{-1}, the pressure drop will be obtained in dynes $\cdot$ cm^{-2} (gm $\cdot$ cm^{-1} $\cdot$ sec^{-2}). To facilitate conversion of the obtained values into conventional units, Table IX lists some conversion factors. Figure 15 plots the viscosity of the three most commonly used carrier gases against temperature.

The evaluation of Eqs. (40)–(43) shows that the pressure drop of a packed column increases rapidly with reduction of the particle

TABLE IX
Conversion Factors of Pressure Units

	atm	dynes $\cdot$ cm^{-2}	psi	gm $\cdot$ cm^{-2}
atm	1	1.01325×10^6	14.696	1033.2
dynes $\cdot$ cm^{-2}	0.98692×10^{-6}	1	1.4504×10^{-5}	1.01971×10^{-3}
psi	0.068046	6.8947×10^4	1	70.307
gm $\cdot$ cm^{-2}	9.6784×10^{-4}	980.665	0.014223	1

---

[17] According to Dal Nogare, the effective particle diameter is about 20% smaller than the average particle diameter calculated from the mesh size.

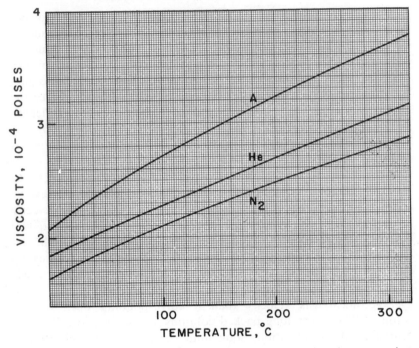

Figure 15. Plot of the viscosities of the three most commonly used carrier gases against temperature.

diameter and increased column length, while in the case of open tubular columns the internal diameter and the length of the column are the dominant factors.

In the next chapter, the equations expressing the pressure drop through the columns are used for the evaluation of actual columns in order to demonstrate the practical limitations which are set by the pressure drop and permeability.

2.333 Comparison of Practical Columns. In the preceding discussions, it was shown that the number of theoretical plates required to separate two adjacent peaks defined by k and α can be calculated. The optimum and practical HETP and gas velocity values were also discussed.

The van Deemter equation describes the correlations between the obtainable HETP values and various column parameters for packed columns. Among these parameters, the particle diameter and the linear gas velocity have the greatest influence, and without going into any further details, it can be stated that the linear gas velocities used with packed columns are of the same order of magnitude as those used with open tubular columns. Also, the HETP values of packed and open tubular columns are of the same order of magnitude. It is also clear that the HETP values are independent on column length if the average linear gas velocity is kept constant. Since the number of theoretical plates of a column is the ratio of column length to HETP, theoretically any plate number could be achieved by any —packed and open tubular—column if a long enough column is used and the proper average linear gas velocity is maintained.

However, there is always a practical limitation, depending primarily on the required pressure drop. The discussion in previous chapters showed that packed columns have a much higher velocity resistance than open tubular columns of similar HETP, and Eq. (43) described the relationship between column permeability, length, and pressure drop. Using the values listed in Table VIII, we can readily establish that, for the same column length, a packed column needs about 100–400 times higher inlet pressures to maintain a given average linear gas velocity than a corresponding open tubular column. Thus, very long packed columns are not practical because instrumental considerations soon limit the possible maximum inlet pressures.

A second factor which has to be taken into consideration is the analysis time. Equation (18) showed the correlation between retention time, column length, average linear gas velocity, and partition ratio. Since the retention time is directly proportional to the column length, a longer column automatically means a longer analysis time. It is, however, often not fully understood that in the case of long open tubular columns, the smaller k value and the possibility of using higher flow rates actually can result in shorter retention times than those for the corresponding, much shorter, packed columns.

In order to demonstrate these variables, let us take three hypothetical columns:

(a) a 3 meter × 2.2 mm ID ($\frac{1}{8}$ in. OD) packed column prepared using a 80/100 mesh support ($\varepsilon = 0.40$, $B_0 = 1.96 \times 10^{-7}$ cm^2);

(b) a 50 meter × 0.25 mm ID open tubular column ($B_0 = 195 \times 10^{-7}$ cm^2);

(c) a 25 meter × 0.25 mm ID open tubular column ($B_0 = 195 \times 10^{-7}$ cm^2).

TABLE X
Comparison of Three Hypothetical Columns

Parameter	Symbol	Dimension	Packed	Open tubular	Open tubular
Column length	L	cm	300	5000	2500
Internal diameter	d	mm	2.2	0.25	0.25
Specific permeability	B_0	$\times 10^{-7}$ cm^2	1.96	195	195
Interparticle porosity	ε	—	0.40	1	1
Flow rate at column outlet	F_c	ml/min	40	1.5	1.14
Outlet linear gas velocity	u_o	cm/sec	43.85	51.75	38.81
Compressibility correction factor	j	—	0.29	0.54	0.72
Inlet pressure	p_i	atm abs.	5.1	2.5	1.75
Outlet pressure	p_o	atm abs.	1	1	1
Pressure drop	Δp	atm	4.1	1.5	0.75
Average linear gas velocity	$\bar{u}$	cm/sec	12.72	27.94	27.94
Relative retention	α		1.10	1.10	1.10
Capacity ratio of second peak	k_2		30	3	3
HETP$_{min}$		mm		0.196	0.196
HETP		mm	0.60	0.588	0.588
Number of theoretical plates	n		5000	85,034	42,517
Resolution	R		1.5	5	3.5
Retention time of second peak	t_{R2}	min	12.19	11.93	5.96
Relative retention	α		1.10	1.10	1.10
Capacity ratio of second peak	k_2		2.50	0.25	0.25
HETP$_{min}$		mm		0.103	0.103
HETP		mm	0.8	0.257	0.257
Number of theoretical plates	n		3750	199,100	97,280
Resolution	R		1	2	1.4
Retention time of second peak	t_{R2}	min	1.38	3.73	1.86

A component pair with a relative retention of $\alpha = 1.10$ is analyzed on these columns; the values of the capacity ratios are $k_P = 30$ and $k_T = 3$. The flow rate (F_c) on the packed and the longer open tubular column is taken as 40 and 1.5 ml/min, respectively; for the shorter open tubular column, the average linear gas velocity $(\bar{u})$ is taken as equal to that of the longer open tubular column, and the values of u_o and F_c are calculated. The HETP of the packed column is taken as 0.6 mm, and a 33% utilization of the available $HETP_{min}$ of the open tubular columns is considered. All these assumptions correspond to average laboratory results, and, in fact, the choice favors the packed column slightly. Finally, let us suppose that the analysis is carried out at 60°C with helium carrier gas; this gives a gas viscosity value of 2.1×10^{-4} poise.

Table X summarizes the assumed values and (down to the double horizontal line) parameters calculated from them by use of the relations discussed in earlier chapters. All the calculations are straightforward except the calculation of Δp, which requires knowledge of j. The combination of Eqs. (16), (17a), and (43) yields a third-order equation. This can be solved, approximately, by assuming a value for Δp and calculating the value of j from this $(p_o = 1 \text{ atm})$. Δp is then calculated using Eq. (43), and if it differs significantly from the assumed value, a better value is arrived at by an iteration of the process.

As seen in Table X, the 50-meter-long open tubular column gives a resolution of $R = 4.97$, while the retention time of the second peak is practically the same as obtained on the packed column, where the two peaks were just completely separated $(R = 1.5)$. On the other hand, working with the 25-meter-long open tubular column, a resolution of $R = 3.5$ can be obtained in half of the analysis time necessary on the packed column.

Based on the data listed in Table X, three further characteristic cases can be evaluated:

(a) If a resolution of $R = 4.97$ is desired on the packed column, then 49,415 theoretical plates would be necessary, which gives a column length of 29.65 meters. A pressure drop of 404 atm through this column would be required to maintain an average linear gas velocity of 12.72 cm/sec, and the retention time of the second peak would be 120.4 min as compared to 11.93 min on the 50-meter-long open tubular column.

(b) If a baseline separation of the two consecutive peaks (characterized by $R = 1.5$) is sufficient, the 50-meter-long open tubular
column can completely separate two components having a
relative retention of $\alpha = 1.03$ while the packed column is only
capable of achieving this resolution for two components with a
relative retention of $\alpha = 1.10$.

(c) On the other hand, if the two components are characterized by
a relative retention of $\alpha = 1.10$ and a baseline separation
($R = 1.5$) is accepted, then only 10,465 plates are necessary on
the open tubular column. This gives a column length of only
6.154 meters. If the same average linear gas velocity is assumed,
the following values will characterize the carrier gas flow
through this column: $\Delta p = 0.18$ atm, $u_o = 30.70$ cm/sec,
$F_c = 0.90$ ml/min. In this case, the retention time of the
second peak is only 1.45 min as compared to 12.2 min on the
packed column.

An additional advantage of open tubular columns can also be
calculated from Table X. When bulk-property detectors (e.g., katha-
rometers) are employed, it is desirable to minimize the dilution of
vapor-pressure-limited samples at the point of injection. Thus, for the
columns characterized in Table X, if the open tubular columns are
just long enough to match the resolution of the packed column, a
tenfold increase in detector response will be realized for such a sample.

The above example demonstrates a case where the capacity ratio
of the open tubular columns is close to optimum.[18] The following
example investigates a case where the k value of the open tubular
column is taken as 0.25 while the k value of the packed column is
set as 2.5. This comparison is definitely unfavorable for the open
tubular column since—as mentioned earlier—with k_T values below
unity, the number of theoretical plates necessary for a given separation
increases significantly.

The lengths, diameters, and flow rates of the three columns and
the relative volatility are all taken equal to those in the previous
comparison. The HETP of the packed column is set as 0.8 mm,
since with packed columns the HETP generally decreases with k and
an HETP of 0.6 mm was originally assumed for $k = 30$. For the open

---

[18] As will be shown in chapter 2.34, the optimum performance of an open tubular
column occurs—according to Golay—at $k = 3.1$.

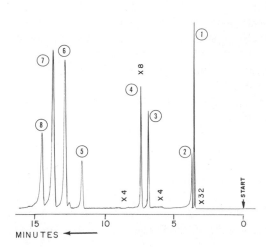

Figure 16. Analysis of fully methylated sugar glycosides, I. [74]. Column: 150 ft × 0.50 mm ID open tubular coated with Apiezon L grease liquid phase. Carrier gas (He) flow rate at outlet and average linear gas velocity: 3.2 ml/min, 25.8 cm/sec. Column temperature: 150°C. Peaks: *1* and *2* solvents (chloroform + pyridine), *3* β- and *4* α-methyl-2,3,4-tri-*O*-methylxylopyranoside, *5* β- and *7* α-methyl-2,3,4,6-tetra-*O*-methylglycopyranoside, *6* α-methyl-2,3,4,6-tetra-*O*-methylmannopyranoside, *8* one of the methyl-2,3,4,6-tetra-*O*-methylgalactopyranoside isomers.

tubular columns, a 40% utilization of the available $\mathrm{HETP_{min}}$ is assumed. Actually this value is much too low. All further calculations are straightforward, and the results are given under the double horizontal line in Table X. As can be seen, a resolution of $k = 1$ is obtained on the packed column and resolutions of 3.2 and 2.2 on the longer and shorter open tubular columns. On the other hand, we could again calculate the necessary length of an open tubular column to obtain a resolution of $R = 1$ for the given peaks. In this case (at $k = 0.25$ and $\alpha = 1.10$), 48,400 theoretical plates are necessary, which corresponds to a column 12.44 meters long. Assuming that the average linear gas velocity remains unchanged, the following values will characterize the carrier gas flow ($p_o = 1$ atm): $\Delta p = 0.37$ atm, $u_o = 34.32$ cm/sec, $F_c = 0.99$ ml/min. The retention time of the second peak in this case is only 0.93 min as compared to 1.38 min on the packed column necessary for the same resolution.

The influence of carrier gas flow rate on efficiency and analysis time is illustrated in Figures 16 and 17 [74], which show the analysis of fully methylated sugar glycosides on the same column but with different flow rates. The average linear gas velocity of the second run is nearly twice that of the first analysis; as a result, the analysis time is reduced to about half. At the same time, however, the number of theoretical plates of peak No. 7 changes from 42,600 down to 23,000.

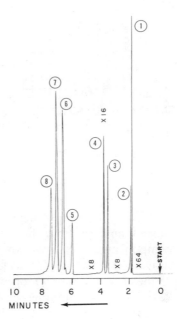

Figure 17. Analysis of fully methylated sugar glycosides, II. [74]. Column, temperature, and peaks as in Fig. 16. Carrier gas (He) flow rate at outlet and average linear gas velocity: 7.4 ml/min, 46.2 cm/sec.

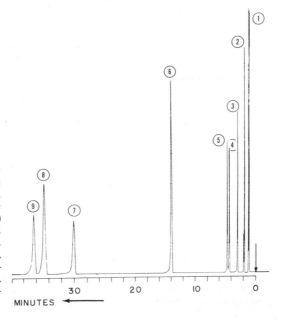

Figure 18. Analysis of a wide-boiling-range aromatic mixture [151]. Column: 25 meters × 0.25 mm ID open tubular, coated with Apiezon L grease liquid phase. Carrier gas (He) flow rate at outlet: 2.3 ml/min. Column temperature: 200°C. Peaks: *1* benzene, *2* indene, *3* naphthalene, *4* β- and *5* α-methylnaphthalene, *6* fluorene, *7* dibenzothiophene, *8* phenanthrene, *9* anthracene.

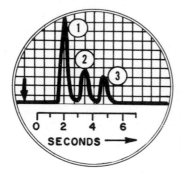

Figure 19. High-speed chromatogram. Column: 16 ft × 0.25 mm ID open tubular coated with Carbowax 1540 polyethylene glycol liquid phase. Carrier gas (He) inlet pressure and flow rate at outlet (calculated): 10 psi (gauge), 55.3 ml/min. Column temperature: 25°C. Oscilloscope readout. Peaks: *1* air, *2* acetone, *3* methylene chloride.

The utilization of higher flow rates not only helps in achieving shorter analysis times but also allows the analysis of high-boiling components at relatively low temperatures. Figure 18, for example, shows the analysis of a wide-boiling-range aromatic mixture on a 25 meter × 0.25 mm ID open tubular column at 200°C [151]; the last component, anthracene (BP = 354°C), emerged in about 37 min.

The combination of a relatively very short open tubular column and high flow rates could result in very fast analysis. In such cases, high-speed recorders or oscilloscope readout have to be utilized. The first chromatograms with oscilloscope readout were obtained by Golay, in the early months of 1957 [112, 215, 216]; Figure 19 reproduces one of his original chromatograms. Figure 20 [55] and Figure 21 [190] show two more high-speed chromatograms; in both cases, high-speed galvanometric recorders were used.

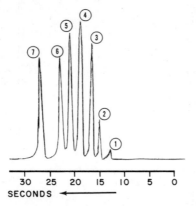

Figure 20. Analysis of C_4 hydrocarbons [55]. Column: 25 ft × 0.25 mm ID open tubular coated with tetraethyleneglycol dimethyl ether liquid phase. Column temperature: − 6°C. Peaks: *1* air, *2* isobutane, *3* *n*-butane, *4* butene-1 + isobutylene, *5* *trans*-butene-2, *6* *cis*-butene-2, *7* butadiene-1,3.

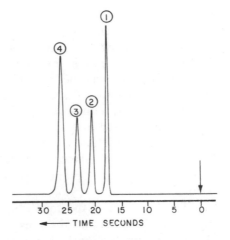

Figure 21. Analysis of butyl alcohols [190]. Column: 30 ft × 0.50 mm ID open tubular, coated with Carbowax 20M polyethyleneglycol liquid phase. Carrier gas (N₂) inlet pressure and flow rate at outlet: 5 psi (gauge), 19 ml/min. Column temperature: 105°C. Peaks: 1 tertiary-, 2 secondary-, 3 iso-, and 4 normal butyl alcohol.

In the last few years, Scott [244, 246, 248] and Desty [65] have demonstrated the separation of multicomponent mixtures in a matter of seconds on systems incorporating high-speed open tubular columns and oscilloscope readout.

2.34 The Performance Index

The discussion in the preceding chapters concerned the influence of column length and carrier gas flow rate on column efficiency and analysis time. It was demonstrated that efficiency and speed can be "traded in." Therefore, it is not the number of theoretical plates that is of primary interest to the practical chromatographer, but rather the rate of production of the theoretical plates. To him, this is the true performance criterion of the column.

In the development of his theory, Golay introduced a useful concept, the performance index (PI), defined by the following formula:

$$PI = \left(\frac{w_h}{t_R - t_M}\right)^4 \left[\left(\frac{t_R - t_M}{t_R}\right)^4 \left(\frac{t_M}{t_R - 0.9375 t_M}\right)\right] t_R \Delta p \qquad (44)$$

In this equation, the retention times and the peak width at half height (w_h) have to be expressed in seconds and the pressure drop across the column (Δp) in dynes·cm^{-2}; thus, the dimension of the performance index will be viscosity, in poises (i.e., dynes·sec·cm^{-2}).

The PI permits a column to be evaluated in terms of readily observable parameters, and assumes only operation at optimum carrier gas velocity. It is numerically equal to a rather large multiple of the carrier gas viscosity.

It is instructive to examine the formula for PI term by term, as it gives a valuable insight into the basic trading rules.

The first term is the reciprocal of the resolving power of the column[19] raised to the fourth power. The smaller this term, the better. The third term is the product of retention time and pressure drop, which should also be minimized. Clearly, for a column of a given performance index, the first and third terms are minimized when the middle (bracketed) term is maximized. This term indicates that the column has optimum performance if the capacity ratio (k) is equal to 3.1.

The last term of the equation, the product of the waiting time (the retention time) and the pressure drop is "the price paid for a certain column performance": it shows how, for a given column, analysis time and pressure may be traded for performance. Thus, for example, a fourfold increase in column length and a fourfold increase in pressure drop (a sixteenfold "price" increase) yield twice the original resolution. Alternatively, the same original resolution could be obtained in one-fourth the original time by halving the length and the radius and operating at four times the pressure drop (and twice the linear velocity). It should be noted that optimum velocity must be maintained when the PI is employed as a trading guide.

The performance index is not a substitute for the plate numbers but an additional measure of column performance; according to Golay's analogy [111], if the number of theoretical plates is likened to the horsepower of a motor, then the PI corresponds to the rate of fuel consumption per unit horsepower.

The PI value of an open tubular column should be of the order of 0.1 poise. With standard packed columns, values of PI have been reported of the order of 1000 poises.[20] With specially built packed columns and in special instrumental systems, values between 50 and 120 have been reported[21]; the best value achieved with such

---

[19] For a definition of this term in accordance with Golay's nomenclature, see chapter 2.353.

[20] M. J. E. Golay, *Nature* **180**:435 (1957).

[21] J. Bohemen and J. H. Purnell, in *Gas Chromatography 1958*, ed. D. H. Desty, Butterworths, London, 1958, pp. 6–22.

a special packed column is given as 5.5 by Scott[22] and, as pointed out by Golay in the discussion of the cited paper, this is about the ultimate performance of a packed column. On the other hand, Condon [54] reported a PI value of 0.87, which is less than a factor of ten from the theoretical limit, for the first commercial open tubular columns.

2.35   Modified Expressions Related to the Plate Number Concept

In studying the differences between packed and open tubular columns, various researchers modified the original expression for the calculation of the number of theoretical plates in order to be able to compare the efficiency of columns of these two types more directly.

2.351   Number of Effective Plates.   In chapter 2.31, the difference in the capacity ratios of packed and open tubular columns was discussed; it was shown that in the latter the value of k will be smaller and, accordingly, the retention time of an inert component (gas hold-up of the column, t_M) becomes relatively larger.

As discussed in the first part, the efficiency of a column can be expressed by the number of theoretical plates (n). Equation (8) can also be written in the following form:

$$n = 16\left(\frac{t'_R + t_M}{w_b}\right)^2 = 5.54\left(\frac{t'_R + t_M}{w_h}\right)^2 \tag{45}$$

where t'_R is the adjusted retention time of the peak and w_b and w_h are the peak width at base ("band intercept") and at half height, respectively.

It is evident that of the two terms in the numerator the gas hold-up time actually does not contribute to the column's efficiency because it only represents the time necessary for the carrier gas to pass through the column. For example, if one were to connect a long empty capillary tube before the actual column, t_M would be larger and thus a higher apparent theoretical plate number value would be obtained although the actual performance of the column remained unchanged.

In order to eliminate this difficulty, Purnell [226] proposed first to use the adjusted retention time (t'_R) instead of the retention time

---

[22] R. P. W. Scott, in *Gas Chromatography 1958*, ed. D. H. Desty, Butterworths, London, 1958, pp. 189–199.

(t_R) in the calculation of column efficiency. Desty *et al.* [65] defined the *number of effective plates* (N) and demonstrated its relationship to the number of theoretical plates (n) calculated by the usual equation:

$$N = 16\left(\frac{t'_R}{w_b}\right)^2 = 5.54\left(\frac{t'_R}{w_h}\right)^2 = n\left(\frac{k}{k+1}\right)^2 \qquad (46)$$

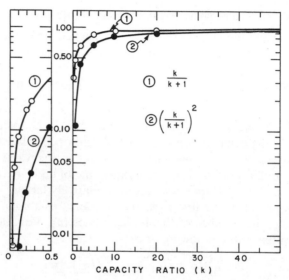

Figure 22. Plot of $[k/(k+1)]$ and $[k/(k+1)]^2$ against the capacity ratio (k). The left side of the diagram is an enlargement of the region up to $k = 0.5$. For numerical values see Table I.

In packed columns, k is usually fairly large; thus, as seen in Figure 22 $[k/(k+1)]^2$ approaches unity and, therefore, $N \simeq n$. With open tubular columns, if k is small, N is only a fraction of n. This means that for early peaks, the very large theoretical number values do not allow *direct* comparison with the performance of a packed column, where the k value for the same peak would be larger. On the other hand, with increasing values of k, N very soon starts to approach the values of n and, for example, at $k = 5$, N is already 69% of n.

Desty *et al.* proposed that in *direct* comparison with packed columns the performance of the open tubular column should be given in effective plate numbers while that of the packed column is given in the usual number of theoretical plates. It should be noted, however, that even for packed columns the k value is often smaller than 5, and in such cases the use of the theoretical plate numbers as compared to effective plate numbers on the open tubular column can be misleading.

The following example compares the values of n and N for an actual chromatogram. Figure 23 plots both values for the chromatogram shown in Figure 5. It can be seen that with increasing k the

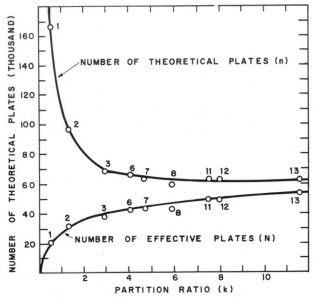

Figure 23. Plot of the number of theoretical plates (n) and effective plates (N) against the capacity ratio (k) for the chromatogram shown in Figure 5. Column, conditions, and corresponding sample components are given there.

two plots level off rapidly and approach each other. At the same time, the plots also show that even for the earliest peaks ($k < 0.3$) $N = 20,000$. Here again we see the superior performance of the open tubular column since—as shown in chapter 2.33—a packed column with such a high theoretical plate number could easily be impractical.

According to Desty *et al.*, the use of the effective plate number concept allows further conclusions to be drawn regarding the true column performance, which could be defined as the *rate of production of effective plates*, i.e., N/t_R. This definition is, in fact, close to the performance index concept. A plot of N/t_R against the linear gas velocity will have a maximum at the optimum practical linear gas velocity (OPGV) value.

2.352 Height Equivalent to One Effective Plate. Equation (9) in the first part of this book introduced the height equivalent to one theoretical plate (HETP) as the ratio of the column length (*L*) to the number of theoretical plates (*n*). A similar expression can also be derived from the number of effective plates; Desty *et al.* called it the *height equivalent to one effective plate* (HEETP):

$$\text{HEETP} = \frac{L}{N} = \text{HETP}\left(\frac{1+k}{k}\right)^2 \tag{47}$$

The basic Golay equation [Eq. (19)] can also be modified in order to express the HEETP; in this case, the three terms (*B*, C_G, and C_L) have to be multiplied by $[k/(k + 1)]^2$.

HEETP *vs.* $\bar{u}$ curves can be drawn, and they look similar to the familiar HETP *vs.* $\bar{u}$ plots. From these curves, the optimum practical gas velocity (OPGV) can be obtained as the point on the curve at which the line drawn to the origin has a minimum slope.

2.353 Modified Expression for the Number of Theoretical Plates. The usual equation for the calculation of the number of theoretical plates is based on the assumption that t_R is relatively much larger than t_M. As seen in chapter 2.351, this is not the case for the early peaks obtained on open tubular columns and, in fact, it is sometimes not even true for packed columns. Therefore, as early as 1956 Golay introduced[23] a modified expression for the calculation of the number of theoretical plates:

$$n_0 = 5.54\left[\frac{t_R}{w_h}\left(\frac{t_R - t_M}{w_h}\right)\right] = 5.54\frac{t_R t_R'}{w_h^2} \tag{48a}$$

Further modification of the right-hand side results in the following equation:

$$n_0 = 5.54\left(\frac{t_R}{w_h}\right)^2\left(\frac{k}{k+1}\right) = n\left(\frac{k}{k+1}\right) \tag{48b}$$

[23] M. J. E. Golay, 129th National Am. Chem. Soc. Meeting, Dallas, Texas, April 1956; *Anal. Chem.* **29**:928 (1957); see also [108].

The difference between Eq. (46) and Eq. (48b) is only that in the latter the $[k/(k + 1)]$ term is not squared.

Golay called the fraction $(t_R - t_M)/w_h$ in Eq. (48a) the *resolving power of the column* and its inverse, $w_h/(t_R - t_M)$, the *relative band width*.

2.354   Resolution Factor. Halász and Schreyer [129], trying to modify the theoretical plate number concept to obtain a more meaningful expression, derived a term very similar to the effective plate number. To obtain this term, Eq. (46) has to be modified by substituting the peak width at the inflection points (w_i) for the peak width at base (w_b). As discussed elsewhere [77], $w_b = 2w_i$, and therefore

$$N = 16\left(\frac{t'_R}{2w_i}\right)^2 = 4\left(\frac{t'_R}{w_i}\right)^2 = 4W^2 \tag{49}$$

Halász and Schreyer called the term W the *resolution factor* and defined its physical meaning as the number of peaks with continuously growing width which can be drawn between the inert gas peak and the peak of interest so that the peaks touch each other at the baseline (i.e., are separated by a resolution of $R = 1.5$). In other words, the resolution factor is the maximum number of components that can be completely resolved up to the peak of interest. The plot of W against k (or t'_R) is characteristic for each column and makes it possible to predict whether a given pair of components can be separated on that column.

If the values of W for a given system are plotted against the column temperature, the curve has a maximum for each component which corresponds to the optimum resolution. This is in good agreement with the results of Scott [247].

2.36   Practical Applications

In the preceding chapters, the advantages of open tubular columns were discussed in some detail. For the practical gas chromatographer, two basic conclusions are particularly important: first, that generally open tubular columns have higher efficiency for the separation of two closely spaced peaks and, second, that resolution and analysis time can be "traded" to advantage.

These two conclusions automatically indicate the two most important fields in which the application of open tubular columns is specially recommended: separation of closely related isomers and analysis of complex mixtures.

The chromatograms used as illustrations in the various chapters of this book were selected to represent a wide variety of problems. Even so, it is obvious that they can illustrate only a fairly small segment of the wide range of possible successful applications of open tubular columns. Therefore, we would like to add ten more chromatograms in which some important applications of common interest are illustrated. These examples demonstrate the analysis of either natural samples or complex synthetic mixtures. In the latter case, an attempt has always been made to select chromatograms showing the separation of "typical samples." However, the author is well aware from his own experience that no such samples exist, and even when a chromatogram apparently shows the analysis of a very complicated sample, practical chromatographers will find some components missing which are important for them.

All of the analyses shown in these chromatograms were carried out under isothermal conditions, which is somewhat unfavorable for wide-boiling-range samples, where programmed-temperature

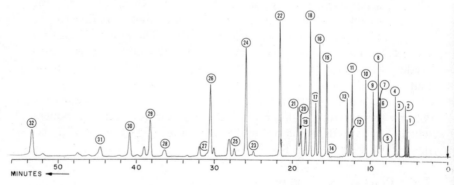

Figure 24. Analysis of a light gasoline [234]. Column: 50 meters × 0.25 mm ID open tubular coated with squalane liquid phase. Column temperature: 40°C. Identified peaks: *1* isobutane, *2* n-butane, *3* isopentane, *4* n-pentane, *5* 2,2-dimethylbutane, *6* cyclopentane, *7* 2,3-dimethylbutane, *8* 2-methylpentane, *9* 3-methylpentane, *10* n-hexane, *11* methylcyclopentane, *12* 2,2-dimethylpentane, *13* benzene, *14* 3,3-dimethylpentane, *15* cyclohexane, *16* 2-methylhexane, *17* 2,3-dimethylpentane, *18* 3-methylhexane, *19* trans-1,3-dimethylcyclopentane, *20* cis-1,3-dimethylcyclopentane + 3-ethylpentane, *21* trans-1,2-dimethylcyclopentane, *22* n-heptane, *23* cis-1,2-dimethylcyclopentane, *24* methylcyclohexane, *25* dimethylhexane, *26* toluene, *27* trans-1,2-cis-4-tetramethylcyclopentane, *28* 2,3-dimethylhexane, *29* 2-methylheptane, *30* 3-methylheptane, *31* cis-1,3-dimethylcyclohexane + trans-1,4-dimethylcyclohexane, *32* n-octane.

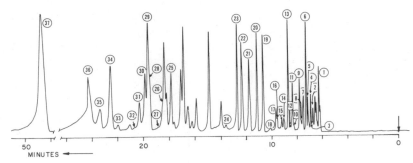

Figure 25. Analysis of a light petroleum oil [234]. Column: 50 meters × 0.25 mm ID open tubular coated with squalane liquid phase. Column temperature: 100°C. Identified peaks: *1* cyclopentadiene, *2* 2-methylpentane, *3* 3-methylpentane, *4* n-hexane, *5* methylcyclopentane, *6* benzene, *7* cyclohexane, *8* 3-ethylpentane, *9* n-heptane, *10* cis-1,2-dimethylcyclopentane, *11* methylcyclohexane, *12* dimethylhexane, *13* toluene, *14* 2,3-dimethylhexane, *15* 2-methylhexane + 3-methylhexane, *16* trans-1,4-dimethylcyclohexane, *17* n-octane, *18* cis-1,4-dimethylcyclohexane, *19* phenylacetylene, *20* ethylbenzene, *21* p/m-xylene, *22* styrene, *23* o-xylene, *24* n-nonane, *25* α-methylstyrene, *26* 1,3,5-trimethylbenzene, *27* tert-butylbenzene, *28* m-methylstyrene, *29* p-methylstyrene, *30* o-methylstyrene, *31* 1,2,4-trimethylbenzene, *32* sec-butylbenzene, *33* n-decane, *34* dicyclopentadiene + β-methylstyrene, *35* 1-methyl-4-isopropylbenzene, *36* indene, *37* naphthalene.

operation is usually preferred. Illustrations of the latter will be given in the fourth part of this book.

One of the largest fields in which open tubular columns are used is the separation of various hydrocarbons. Their natural mixtures are either of petroleum or coal tar origin. Figures 24 and 25 illustrate the analysis of two petroleum fractions [234]. These chromatograms are taken from a paper by Schneck, describing the routine control of a cracking plant with help of open tubular columns. Figure 26 shows the analysis of some olefinic hydrocarbons [166]; here the unidentified peaks correspond to some further isomers in this range.

Coal tar is particularly rich in a large variety of aromatic substances. Figure 5 already demonstrated the analysis of a tar fraction consisting of the benzene homologs. Figure 27 now shows the analysis of a cut containing the naphthalene homologs [159]. Here, except for the 2,6- and 2,7-dimethylnaphthalene isomers, all the methyl-, ethyl-, and dimethylnaphthalenes are separated. The unidentified peaks probably correspond to ethyl- or dimethylthionaphthenes and peaks *20–25* to trimethylnaphthalenes.

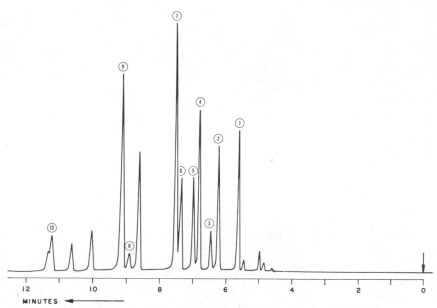

Figure 26. Analysis of $C_5 - C_6$ olefins [166]. Column: 50 meters × 0.25 mm ID open tubular coated with di-2-ethylhexylsebacate liquid phase. Carrier gas (He) flow at outlet: 1.25 ml/min. Column temperature: 25°C. Identified peaks: *1* 3-methylbutene-1, *2* *n*-pentane + pentene-1, *3* 2-methylbutene-1, *4* *trans*-pentene-2, *5* *cis*-pentene-2, *6* 2-methylbutene-2, *7* 2,2-dimethylbutane, *8* *trans*-4-methylpentene-2, *9* *cis*-4-methyl-pentene-2, *10* hexene-1.

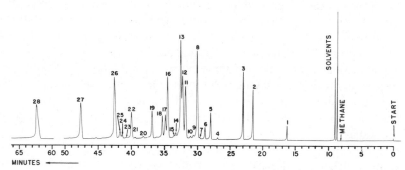

Figure 27. Analysis of the dimethylnaphthalene fraction of a bituminous coal tar [159]. Column: 300 ft × 0.25 mm ID open tubular coated with DC-550 phenyl silicone oil liquid phase. Carrier gas (He) inlet pressure: 30 psi (gauge). Column temperature: 160°C. Identified peaks: *1* naphthalene, *2* 2-methylnaphthalene, *3* 1-methylnaphthalene, *5* biphenyl, *6* 2-ethylnaphthalene, *7* 1-ethylnaphthalene, *8* 2,6-dimethylnaphthalene + 2,7-dimethylnaphthalene, *11* 1,7-dimethylnaphthalene, *12* 1,3-dimethylnaphthalene, *13* 1,6-dimethylnaphthalene, *16* 1,5-dimethylnaphtha-lene, *17* 1,4-dimethylnaphthalene, *18* 2,3-dimethylnaphthalene, *19* 1,2-dimethyl-naphthalene, *26* acenaphthene, *27* diphenylene oxide, *28* fluorene.

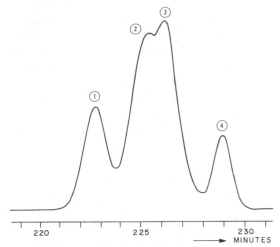

Figure 28. Separation of benzene and deuterobenzenes [29]. Column: 250 meters × 0.28 mm ID open
tubular coated with squalane liquid phase. Carrier
gas (N_2) flow rate at outlet: 0.25 ml/min. Column
temperature: 10°C. Peaks: *1* C_6H_6, *2* $C_6H_3D_3$, *3*
$C_6H_2D_4$, *4* C_6D_6.

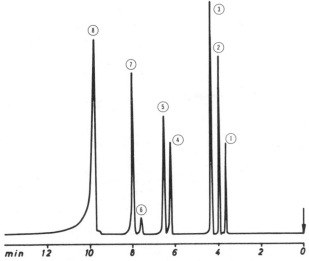

Figure 29. Analysis of substituted aromatics [166]. Column:
50 meters × 0.25 mm ID open tubular coated with FS-1265
fluorinated silicone oil liquid phase. Column temperature: 150°C.
Peaks: *1* benzene, *2* toluene, *3* ethylbenzene, *4* *o*-bromotoluene,
5 *p*-bromotoluene, *6* *o*-bromo-*m*-xylene, *7* *p*-bromo-*m*-xylene,
8 nitrobenzene.

Open tubular columns allow not only the separation of closely related isomers but also that of isotopically substituted isomers. Falconer and Cvetanovic [90], for example, readily separated some isopentane and 2,3-dimethylbutane isomers: $C_5H_2D_{10}$, $C_5H_6D_6$, C_5H_{12}, and $C_6H_2D_{12}$, $C_6H_8D_6$, C_6H_{14}. Bruner *et al.* [29] showed the separation of benzene and deuterobenzenes. As seen in Figure 28, a complete separation of C_6H_6 and C_6D_6 and a partial separation of $C_6H_3D_3$ and $C_6H_2D_4$ were obtained using a long squalane column.

Figure 29 demonstrates the separation of some brominated, alkyl-substituted benzene homologs [166]. These substances are polar in nature and at the same time are closely related isomers. The sample also contained nitrobenzene, and its peak shows definite tailing; this is due to the highly polar nature of this substance. Another complicated halogenated aromatic sample consisting of

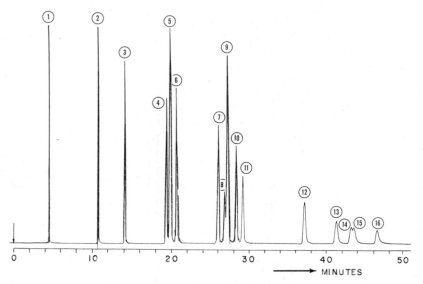

Figure 30. Analysis of chlorinated biphenyls [279]. Column: 200 ft × 0.25 mm ID open tubular coated with Apiezon L grease liquid phase. Carrier gas (A) inlet pressure: 30 psi (gauge). Column temperature: 185°C. Peaks: *1* benzene (solvent), *2* biphenyl, *3* 2-chlorobiphenyl, *4* 2,2'-dichlorobiphenyl, *5* 3-chlorobiphenyl + 2,6-dichlorobiphenyl, *6* 4-chlorobiphenyl, *7* 2,5-dichlorobiphenyl, *8* 2,4-dichlorobiphenyl, *9* 2,3'-dichlorobiphenyl, *10* 2,3-dichlorobiphenyl, *11* 2,4'-dichlorobiphenyl, *12* 3,5-dichlorobiphenyl, *13* 3,3'-dichlorobiphenyl, *14* 3,4-dichlorobiphenyl, *15* 3,4'-dichlorobiphenyl, *16* 4,4'-dichlorobiphenyl.

the isomeric chloro- and dichlorobiphenyls was analyzed by Wein-garten *et al.* [279]. Figure 30 shows the analysis of this mixture on a column coated with Apiezon L liquid phase.

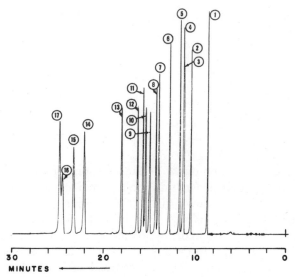

Figure 31. Analysis of anisole homologs [256a]. Column: 50 meters × 0.25 mm ID open tubular coated with 2,4-xylenyl phosphate liquid phase. Column temperature: 140°C. Peaks: *1* anisole, *2* o-methylanisole, *3* m-methyl-anisole, *4* p-methylanisole, *5* 2,6-dimethylanisole, *6* o-ethylanisole, *7* 2,5-dimethylanisole, *8* 2,3-dimethylanisole, *9* m-ethylanisole, *10* p-ethylanisole, *11* 3,5-dimethylanisole, *12* 2,4-dimethylanisole, *13* 3,4-dimethylanisole, *14* o-dimethoxybenzene, *15* 2,3,5-trimethylanisole, *16* p-dimeth-oxybenzene, *17* m-dimethoxybenzene.

Excellent separation of the isomeric anisole homologs is shown in Figure 31 [256a]. Except for the *m*- and *p*-methylanisole pair, for which only a starting separation is indicated, all the methyl-, ethyl-, and dimethylmethoxybenzenes and the three dimethoxybenzenes are well separated.

Figure 32 shows the separation of C_1–C_5 alcohols.[24] These substances are highly polar and their analysis represents a problem

---

[24] Chromatogram reproduced from data sheet F-9791 of the Barber–Colman Company, Rockford, Ill.

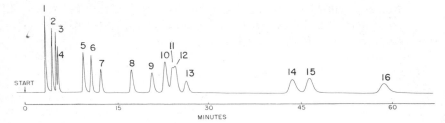

Figure 32. Analysis of C_1–C_5 alcohols[24]. Column: 100 ft × 0.25 mm ID open tubular coated with Armeen SD liquid phase. Column temperature: 60°C. Carrier gas (A) inlet pressure: 15 psi (gauge). Peaks: *1* methanol, *2* ethanol, *3* isopropanol, *4* tert-butanol, *5* n-propanol, *6* sec-butanol, *7* 2-methylbutanol-2, *8* isobutanol, *9* 2-methyl-butanol-3, *10* n-butanol, *11* pentanol-3, *12* pentanol-2, *13* 2,2-dimethylpropanol-1, *14* methylbutanol-1, *15* 2-methylbutanol-1, *16* pentanol-1.

even on packed column with Teflon support. Here, Armeen SD was used as the liquid phase; this is a commercial soybean oil amine whose main components are miristyl (1 %), palmityl (24 %), palmitoleyl (1 %), stearyl (8 %), oleyl (49 %), and linoleyl (15 %) amines. As seen on the chromatogram, practically all the isomers are well separated.

The last example is from the field of fatty acid analysis. Natural fatty acid (methyl ester) mixtures obtained after the hydrolization (transesterification) of triglycerides are very complex. Prior to the introduction of gas chromatography, the analysis of the various closely related isomers and homologs was extremely laborious. Open

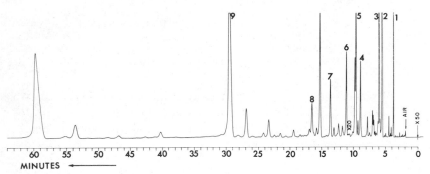

Figure 33. Analysis of the fatty acid methyl ester mixture obtained from menhaden oil[25]. Column: 150 ft × 0.25 mm ID open tubular, coated with butanediol succinate liquid phase. Carrier gas (N_2) inlet pressure: 40 psi (gauge). Column temperature: 185°C. Sample size: 1 μl (liquid), split. Identified peaks: methyl esters of *1* myristic, *2* palmitic, *3* palmitoleic, *4* stearic, *5* oleic, *6* linoleic, *7* linolenic (?), *8* gadolic (?), *9* behenic (?) acids.

tubular columns are particularly suitable for such analyses, and, e.g., the publications of Lipsky and Landowne [175, 176, 181, 182] and Litchfield *et al.* [183–186] clearly demonstrated the superiority of such systems for the analysis of wide-boiling-range samples in which closely related isomers and homologs are present. Some examples of the analysis of fatty acid methyl ester mixtures have already been given; Figure 33 now shows the result of the analysis of a menhaden oil methyl ester sample.[25] This chromatogram indicates about 80 components, a few of which are identified. The identification was performed partly by comparison with a standard fatty acid methyl ester mixture, partly by plotting the logarithm of the adjusted retention times against the carbon number and the number of double bonds; since the plotting cannot be taken as absolute proof, peaks from this identification are marked with a question mark.

2.4 THE INFLUENCE OF TEMPERATURE

Working at lower temperatures will increase the value of the partition coefficient. Since the change in the β value is negligible, this will result in larger k values. Since in Eq. (10), the formula for the number of theoretical plates necessary for a given separation, the $[(k + 1)/k]^2$ term has a significant influence in the case of open tubular columns, lowering of the temperature will reduce the number of theoretical plates required for a given separation.

Similar considerations apply to the relative retention. Its temperature dependence can be described by the following equation:

$$\log \alpha = a \frac{1}{T_c} + b \tag{50}$$

where T_c is the column temperature (in °K), and a and b are constants. The plot of $\log \alpha$ against $1/T_c$ may have two forms. In the first case, when $a \simeq 0$, the plot will be practically parallel to the abscissa. This means that the relative retention does not change with temperature. Scott [247] has pointed out that this is the case with a large number of substances of similar type. In the second case, when $a > 0$, the relative retention will increase with decreasing temperature.

Schreyer [239] has investigated a number of plots measured on a squalane column. The first case was found to apply for the pairs

[25] F. J. Kabot (Perkin–Elmer Corp., Norwalk, Conn.) unpublished result. This chromatogram is not identical to that reproduced in part in Figure 14.

benzene–n-hexane, cyclohexane–n-hexane, toluene–m/p-xylene, m-xylene–p-xylene, and n-pentane–isopentane, while for the n-heptane–n-hexane, toluene–n-hexane, and toluene–benzene pairs the value of a was greater than zero.

If $a \simeq 0$, the $[\alpha/(\alpha - 1)]^2$ term in Eq. (10) remains unchanged when the column temperature is reduced. Since, however, the value of the last term decreases, n_{req} will also decrease. In the second case ($a > 0$), the relative retention will increase when the temperature is lowered, resulting in a decrease in the value of the $[\alpha/(\alpha - 1)]^2$ term; this fact contributes to an even greater reduction in the required number of theoretical plates.

The following example demonstrates the influence of column temperature on n_{req}. According to Desty and Goldup [63], the partition coefficient of n-pentane at 25 and 50°C is 123 and 60, respectively. A 0.243 mm ID squalane column is used with a film thickness of 0.55 μ; this gives a β value of 110.45. Thus, the k values calculated from Eq. (4) are 1.11 (25°C) and 0.54 (50°C). Let us suppose a second peak before n-pentane, with a relative retention of $\alpha = 1.05$, and that the relative retention does not change with temperature. Using Eq. (10), we find the following theoretical plate numbers necessary to achieve a resolution of $R = 1.5$:

$$\text{at } 25°C: \qquad n_{req} = 57,300$$

$$\text{at } 50°C: \qquad n_{req} = 128,900$$

These facts would seem to indicate that it is best to operate the column at the lowest possible temperature. In cases where due to its very low capacity ratio (k) value a column does not have high enough efficiency for a desired separation (*cf.* Figure 11), lowering of the temperature will certainly help. However, in general, *the lowest possible temperature is not the right temperature for the analysis.* There are two basic reasons for this.

First, the time effect cannot be neglected. Since the partition ratio will be larger at lower temperatures, the analysis time will also be longer. For example, in the above case—assuming the same average linear gas velocity in both analyses—the retention time of n-pentane at 25°C will be 37% longer than that at 50°C.

The second reason has to do with the overall efficiency of the column. It has been shown earlier that for work above the OPGV the B term in the Golay equation is negligibly small and the HETP is governed by the $(C_G + C_L)$ term. Naturally, one would like to achieve a very small value for this term. The value of $(C_G + C_L)$ will fall with increasing temperature; this means that an open tubular column has generally higher efficiency at higher temperatures. Raising the temperature will also decrease k and, thus, the time of analysis; on the other hand, more theoretical plates will be required for a given separation. This indicates that the selection of proper analysis temperature is a complex question and usually requires a compromise between the desired analysis time and the necessary and available theoretical plates.

Scott [247] investigated the influence of temperature on efficiency, resolution, and analysis time of open tubular columns in detail. He demonstrated that there is always an optimum temperature at which the best resolution can be obtained in the shortest analysis time. He also described certain equations for the theoretical calculation of the optimum conditions; however, these equations are beyond the scope of a practical gas chromatographer's manual. Besides, experimental values showed relatively significant deviations from the calculated values, although they proved the general validity of the rule.

2.5 SAMPLE CAPACITY

It is evident that each gas chromatographic column has a certain sample capacity depending on the amount of liquid phase present. When samples larger than the allowed are introduced, proper equilibrium cannot be established in the various segments (the "theoretical plates") of the column. As a result, the efficiency of the column will be reduced and peak broadening and asymmetry will be observed.

The maximum permissible sample size was defined by Keulemans[26] as the maximum amount which can be injected into a column without more than 10% loss in efficiency. It is expressed[26,27] as follows:

$$v_K = a_K v_{\text{eff}} \sqrt{n} \tag{51}$$

where v_K is the volume of vaporized sample exclusive of carrier gas, v_{eff} is the effective volume of one theoretical plate, n is the number of theoretical plates, and a_K is a constant.

[26] A. I. M. Keulemans, *Gas Chromatography*, 2nd edition. Reinhold, New York, 1959, pp. 124 and 194.

[27] A. Klinkenberg, in *Gas Chromatography 1960*, ed. R. P. W. Scott, Butterworths, London, 1960, pp. 182–183.

By definition, the effective volume of one plate is expressed by Eq. (52):

$$v_{eff} = v_G + K v_L = \frac{V_G}{n} + K\frac{V_L}{n} \tag{52}$$

where v_G and v_L are the gas and liquid volumes per plate while V_G and V_L are the volumes of the gas and liquid phases in the whole column. By substituting $(V_G/V_L)k$ for the partition coefficient, Eq. (52) can be modified:

$$v_{eff} = v_G(1 + k) = \frac{V_G}{n}(1 + k) \tag{53}$$

In practice, V_G may be replaced by V, the geometric volume of the column. Therefore,

$$v_K = \frac{a_K V}{\sqrt{n}}(1 + k) = a_K \frac{r^2 \pi L}{\sqrt{n}}(1 + k) \tag{54}$$

In this equation, a_K is the limiting value for a given system. According to Keulemans, for packed columns a_K is approximately equal to 0.02. However, very few numerical values of v_K and a_K are available for open tubular columns; Table XI summarizes the results of three published measurements. The liquid phase in all three columns was squalane. These values are well substantiated by the data published by Zlatkis and Walker [286]: the maximum sample size resulting in not more than 10% loss in efficiency of a 111-meter-long column with 375 μ radius (coated with hexadecane) was about 3.3×10^{-2} μl liquid, using benzene, cyclohexane, and n-hexane samples.

The "maximum permissible sample size" relates the permitted sample amount to column efficiency. There are, however, two additional values important for the practical analyst: *the maximum sample sizes for which the peak height and the peak area are still linear.* The first of these values represents the linear range of the column, while the second is related mainly to the linear range of the detector. Generally, one can state that when the peak height is no longer linearly proportional to the sample size, the column is overloaded, while nonlinearity in the peak area values indicates an overloaded detector.

TABLE XI
Numerical Values for Maximum Sample Size for Four Different Columns

Ref.	Column length, m	Column radius, μ	u, cm/sec	Sample	n	k	v_k vapor, μl	v_k liquid, μl	a_k
[47]	1.05	17.25	100.0	n-heptane	13,950	2	1.12×10^{-5}	7.31×10^{-8}	4.30×10^{-4}
[47]	273.60	76.00	10.5	n-heptane	712,000	3	2.24×10^{-3}	1.46×10^{-5}	0.95×10^{-2}
[65]	91.44	775.00	53.0	n-hexane	72,000	0.093	73	0.4	104

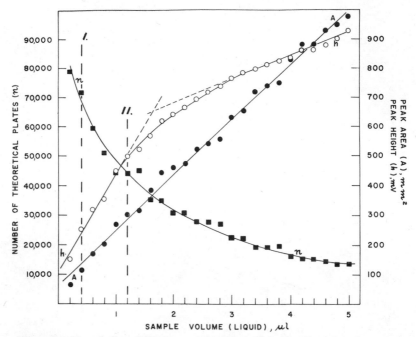

Figure 34. Plot of theoretical plate number (*n*), peak height (*h*), and peak area (*A*)
values against sample volume for a 300 ft × 1.55 mm ID open tubular column
coated with squalane liquid phase, operated at 50°C and approximately 60 ml/min
(He) outlet flow, using a thermistor detector. Sample: *n*-hexane [83].

Figure 34 demonstrates the limitations in the sample size [83].
Here, a 300 ft × 1.55 mm ID open tubular column coated with
squalane was operated at 50°C and with approximately 60 ml/min
helium outlet flow. A thermistor detector was used and samples of
normal hexane (99+ mole percent) were injected into the instrument;
the sample size was varied from 0.2 μl up to 5.0 μl, with 0.2-μl changes
throughout the range. *I* represents the maximum permissible sample
size according to Keulemans' definition (v_K); it is 0.4 μl (liquid) and
is listed in the last row of Table XI. *II* defines the end of the linear
range of the column; it is 1.2 μl (liquid) sample. The peak area,
however, is linear up to at least 5 μl (liquid) sample. The importance
of these two additional sample limits will be discussed in the next
chapter.

In this example, a large-diameter open tubular column was used
because its sample capacity is larger, and therefore such

measurements—with direct sample injection—can be carried out more easily. Similar graphs could also be plotted for columns with smaller diameter; however, the limiting values are naturally much smaller. Therefore, with such columns the direct introduction of the necessary small sample volumes with the usual techniques is practically impossible, and therefore indirect sampling procedures are generally used. These methods are discussed later, in the fourth part of this book.

2.6   OPEN TUBULAR COLUMNS OF VARIOUS DIAMETERS

Equation (29) showed that the theoretically obtainable minimum HETP value is directly proportional to the radius of the column tubing, and it was shown in chapter 2.22 that at the optimum practical gas velocity (OPGV) the HETP obtained is actually proportional to the square of the column radius. Thus, from both points of view, tubes of the smallest possible diameter should be preferred. For practical reasons, however, tubes with an internal diameter of less than 0.25 mm are only rarely used, although Desty et al. [63, 66] demonstrated the excellent performance of such columns. The problem with such very small-ID columns concerns mainly the quality of the available tubing, the necessarily very thin liquid film, and the sample capacity of the column.

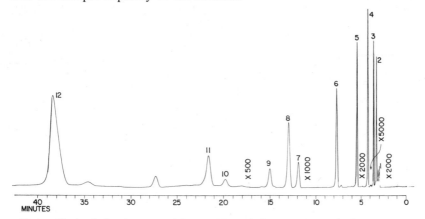

Figure 35. Analysis of a natural fatty acid methyl ester mixture [82]. Column: 200 ft × 0.50 mm ID open tubular coated with butanediol succinate liquid phase. Carrier gas (He) inlet pressure: 12 psi (gauge). Column temperature: 180°C. Sample size: 0.05 μl (liquid), no split. Peaks: methyl *1* caproate, *2* caprylate, *3* caprate, *4* laurate, *5* myristate, *6* palmitate, *7* stearate, *8* oleate, *9* linoleate, *10* linolenate, *11* arachidate, and *12* erucate.

Today, the most frequently used open tubular columns have inside diameters of about 0.25 mm (0.010 in.). However, columns with larger diameters often have many advantages in special, practical applications.

The most significant advantage of such columns is their increased sample capacity. The practical sample capacity of 0.25 mm ID columns is usually of the order of 10^{-3} μl (liquid), and therefore only indirect sample introduction with large split ratios is possible. By only doubling the diameter, the necessary split ratio can be reduced significantly and, depending on the sample, direct sample injection is sometimes made possible. Figure 35, for example, shows the analysis of a natural fatty acid methyl ester mixture on a 200 ft × 0.50 mm ID column [82]. Here, 0.05 μl liquid sample was injected directly, without any split, by a Hamilton syringe of 0.5 μl capacity. Although the peaks are slightly broader than good column performance would show, the separation is still adequate ($R = 1.89$) for the stearate–oleate pair. (The asymmetry of the methyl erucate peak is due to the temperature being too low, rather than to column overloading.)

This increase in sample capacity is even more important if the maximum sample size is not defined according to the (theoretically correct) definition of Keulemans. Jentzsch and Hövermann [153], for example, using a 100 meter × 0.50 mm ID column coated with squalane ($d_f = 1.1$ μ) at 68°C, with a helium flow rate of 21.2 ml/min, and analyzing a mixture of pentane isomers, showed that the number of theoretical plates for the n-pentane peak changed from 20,680 to

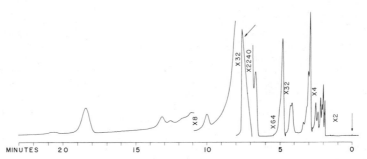

Figure 36. Quality control of n-butyl acetate [152]. Column: 100 meters × 1.0 mm ID open tubular coated with Ucon oil 550-X liquid phase. Carrier gas (He) flow rate at outlet: 37.3 ml/min. Column temperature: 80°C. Sample size: 1 μl (liquid) n-butyl acetate; no split. Peak marked with an arrow corresponds to n-butyl acetate; the others represent impurities present. Flame ionization detector with 2.5 millivolt recorder.

13,100 (i.e., by only 37%) when the amount of *n*-pentane injected was increased tenfold, from 0.056 μl to 0.56 μl (liquid).

Of course, for still larger diameters, the sample capacity increases even more, allowing trace analysis with open tubular columns. This possibility can be further enhanced by permitting overloading of the column by the matrix sample component, as long as the trace components of interest remain in the linear range. This is illustrated in Figure 36, which shows the analysis of a normal butyl acetate sample [152] on a 1 mm ID open tubular column, using a flame ionization detector. Here, the *n*-butyl acetate peak is overloaded on the column; the impurity peaks correspond to concentrations of less than 100 parts per million.

A basic problem with open tubular columns of small diameter is that owing to the very small sample sizes only ionization detectors can be used.[28] With greater diameter and, thus, sample capacity of the column, the criterion for the detector is much less stringent, and conventional thermal conductivity detectors are also satisfactory. The first such chromatograms (using 2 mm ID open tubular columns) were shown by E. Kováts at the meeting of the English Gas Chromatography Discussion Group, Birmingham, April 21, 1961; since then, the literature has reported the application of such systems to numerous problems [152, 155, 228, 241, 259, 268]. Figure 37, for example, shows the analysis of an ester sample on a 300 ft × 1.55 mm ID open tubular column using a commercial thermistor detector. In this chromatogram, the good peak symmetry and column performance despite the fairly high carrier gas flow is noteworthy: about 20–30% of the theoretical optimum column performance was utilized.

Another advantage of open tubular columns with diameters of 0.5 mm or higher is that they allow an even greater flexibility in the selection of carrier gas flow. Higher flow rates can significantly speed up the analysis, and although—as shown earlier—the overall efficiency is reduced, it is usually still adequate for the particular separation. A further gain when using high flow rates is that they allow the analysis of samples well below their boiling points. The most extreme case for such an application was shown by Quiram [228], who used a 250 ft × 1.55 mm ID open tubular column with a

---

[28] It should be emphasized that the real problem is not the "sensitivity" of the thermal conductivity detectors but the much too large volume of commercial models and, sometimes, their slow response time. The problems associated with the detectors used with open tubular columns are discussed in the fourth part of this book.

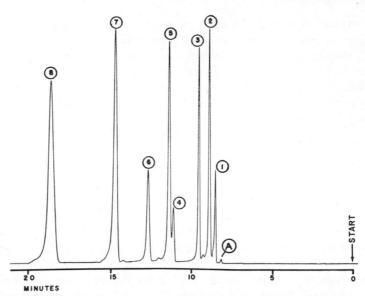

Figure 37. Analysis of an ester sample [83]. Column: 300 ft × 1.55 mm
ID open tubular coated with Ucon oil 550-X liquid phase. Carrier gas
(He) flow rate at column outlet: 25.0 ml/min. Column temperature: 75°C.
Sample size: 1 μl (liquid); no split. Peaks: *A* air, *1* methyl formate,
2 ethyl formate, *3* ethyl acetate, *4* n-propyl acetate, *5* methyl butyrate,
6 isobutyl acetate, *7* n-butyl acetate, *8* n-amyl acetate.

flow rate of 800 ml/min, injecting—without split—7-μl liquid sample
consisting of C_6–C_{16} normal alcohols. As a result, he could analyze
n-cetyl alcohol (BP = 344°C) at 175°C in 8 min (Figure 38).

Finally, a particular advantage of larger-diameter open tubular
columns is that since their volume is larger, the requirements on

Figure 38. Analysis of higher
normal alcohols [228]. Column:
250 ft × 1.55 mm ID open tubular
coated with Surfonic ID 300 (a
tridecyl alcohol + ethylene oxide
condensate; one mole TDA + 30
mole EtOx) liquid phase. Carrier
gas (N_2) flow rate at outlet: 800
ml/min. Column temperature:
175°C. Sample size: 7 μl (liquid);
no split. Peaks: *1* hexyl, *2* octyl,
3 decyl, *4* dodecyl, *5* tetradecyl,
and *6* hexadecyl alcohol.

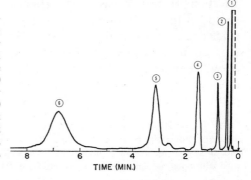

instrument construction are less stringent. When using, e.g., columns of 0.25 mm ID, utmost care must be taken to keep the volumes of the connecting lines between the injection port and the column and between the column and the detector at a minimum. Since the volume of a 150 ft × 0.25 mm ID column is only 2.24 ml and very low flow rates (0.5–3 ml/min) are used, even small dead volumes can result in significant peak spreading and tailing. Consequently, the column often shows an apparent poor performance, although it is the system that is inadequate, and not the column. The volume of columns with larger diameters increases rapidly: the volume of a 150 ft × 0.50 mm ID column is 9 ml and that of a 300 ft × 1.55 mm ID column is as high as 172.4 ml. With such columns, the dead volumes of the connections are much less critical, particularly since the flow rates used are also higher.

Of course, open tubular columns with larger diameters have some disadvantages compared with similar columns with smaller diameters. The problem logically follows the discussion in chapters 2.2–2.4 and is related to the much larger β values of these columns. Thus, the plate number necessary for a given separation may exceed the number of theoretical plates actually available. But since the larger columns have practically no flow resistance, a much longer column can be used and the overall performance increased in this way. In practice, proper balance of column length, diameter, and flow rate will result, in most cases, in a performance superior to that of packed columns.

Third Part

Preparation of Open Tubular Columns

3.1 TUBING

For the preparation of open tubular columns, long tubing of various diameters is used. The characteristics of tubing suitable for constructing such columns for gas chromatography can be summarized as follows:

(a) It should be possible to obtain the tubing in the proper dimensions (length, ID);
(b) The inside diameter should be constant over the entire length of the tube;
(c) The inside surface of the tubing should be uniform, without pores or fissures;
(d) The material of the tubing should not react with the stationary phase or sample components, and should not exhibit significant secondary adsorption power to the sample components;
(e) The tubing should permit the formation of a uniformly stable coating of the inside surface with the stationary phase;
(f) The tubing should have sufficient mechanical strength for use in a gas chromatographic equipment.

Plastic, glass, or metal tubing had been used for the preparation of open tubular columns. Actually, none of the materials available is ideal for column preparation, and each has some drawbacks with respect to the criteria listed above. In practice, metal—particularly stainless steel—tubing is the most widely used.

3.11 Plastic Tubing

The first "open tubular columns" made by Golay consisted of *Tygon tubing* (see chapter 1.1), but such tubes have many disadvantages in practical use. They do not have the necessary mechanical strength and their temperature range is severely limited. The most serious problem, however, is that when these tubes remain in use for a longer period of time, the plasticizer used in the manufacture of the tubing tends to migrate and evaporate.

Nylon tubing was used successfully by Scott [243, 249]. It performs well at lower temperatures and with nonpolar liquid phases, but even at slightly elevated temperatures, unsatisfactory operation has been reported.[29] The problem lies mainly in softening of the tube, solution of the sample components in the tubing material, and permeability for both sample components and atmospheric moisture.

Teflon tubing had also been tried for open tubular columns, but with little success. The basic problem here is that it does not hold the liquid phase on the inside surface.

3.12   Glass Tubing

Glass "capillary tubes" were first used by Desty and co-workers for the preparation of open tubular columns. They also developed a relatively simple machine which permits the drawing of tubes with a wide range of diameters and wall thicknesses, in lengths of several hundred feet, coiled in the form of a compact helix [67]. A similar machine was also described almost simultaneously by Kreyenbuhl [174].

Glass columns have been successfully used with nonpolar phases and some silicone oils [31]. However, the lifetime of such columns seems to be limited: Halász, for example, pointed out[30] that the lifetime of glass columns coated with squalane was only 2–3 days, while copper columns coated with the same liquid phase lasted for at least 7 months. The basic problem here is the insufficient wetting of the inner glass surface by the liquid phase.

Other problems with glass columns concern their fragility and the relative difficulty of connecting them to the other parts of the instrument.

Recently a different application of glass columns has been reported in which the inner surface is etched by some reagents forming

---

[29] Panel discussion in *Progress in Industrial Gas Chromatography*, ed. H. A. Szymanski, Plenum Press, New York, 1961, pp. 229–230.

[30] Panel discussion, in *Gas Chromatography*, ed. N. Brenner, J. E. Callen, and M. D. Weiss, Academic Press, New York, 1961, pp. 521–526, 557–562.

an adsorption layer [30, 171, 200]. This tubing can be used either in gas adsorption chromatography or by coating it with a suitable liquid phase. In the latter case, the roughened interior surface secures better stability of the liquid coating. The methods of preparation of such columns will be discussed in chapter 3.41.

3.13   Metal Tubing

Aluminum, nickel, gold, copper, and stainless steel tubing have been successfully used for the preparation of open tubular columns.

The use of *aluminum tubing* was reported by Kaiser [160], who emphasized the stability of the coating in such columns. However, according to Petitjean and Leftault [217], the natural oxide film which always forms on the surface of aluminum may have significant surface activity, resulting in secondary adsorption and peak tailing.

Nickel tubes can also result in good stable columns. However, the principal problem with them is that they are not readily available in the proper sizes [282].

Gold capillary tubes were used by Desty *et al.* [66]. They showed very good performance; however, they are obviously too expensive for routine use.

Copper columns have been used by many workers [e.g., 66, 282] with good results. They are easy to coat, and for most types of samples they perform properly when used at not too high temperatures. However, the application of such columns at higher temperatures (above about 150–175°C) can result in the decomposition of both the liquid phase and the sample components. For example Kaiser [160] reports that cyclohexanone will decompose at 180°C, and according to Desty *et al.* [66], Apiezon L liquid phase rapidly decomposes at 200°C if air is accidentally admitted to the column.

The most frequently used tubing for the preparation of open tubular columns is made of *stainless steel*. These tubes are available in a wide range of internal and external diameters and in lengths up to 2000 ft or longer. They can be cleaned and coated relatively easily, and generally provide highly durable coatings. They are relatively inert, although in the case of highly polar samples the use of special additives is recommended (see chapter 3.24).

Special stainless steel tubing suitable for the preparation of open tubular columns is marketed by at least three companies in the USA (J. Bishop & Co., Malvern, Pa., Posen & Kline, Norristown,

TABLE XII

Typical Chemical Composition (in %) of Various Stainless Steels Used as Column Tubing

Material	C max	Mn max	Si max	P max	S max	Cr	Ni	Fe	Zn max	Sn max	Pb max	Cu	Other elements
304	0.08	2.00	1.00	0.045	0.030	18–20.00	8–12.00	Bal.	—	—	—	—	—
316	0.08	2.00	1.00	0.045	0.030	16–18.00	10–14.00	Bal.	—	—	—	—	Mo: 2–3
321	0.08	2.00	1.00	0.045	0.030	17–19.00	9–12.00	Bal.	—	—	—	—	Ti: 5 × C
347	0.08	2.00	1.00	0.045	0.030	17–19.00	9–13.00	Bal.	—	—	—	—	Cb & Ti: 10 × C min
3% Cupronickel	—	1.00	—	—	—	—	29–32.00	0.40–0.70	1.00	1.00	0.05	Bal.	—

Pa., and Superior Tube, Norristown, Pa.). These tubes are made mostly of the austenitic chromium nickel stainless steels, AISI Types 304, 316, 321, or 347, or of a 30% cupronickel-type material. Table XII gives typical chemical compositions of these stainless steels.[31]

Since stainless steel has a relatively poor heat conductivity, the selection of the proper wall thickness is important. Generally, there are two types of "capillary" tubing available. In the first, the wall thickness is of the order of 0.12–0.15 mm; these tubes are relatively fragile and therefore can be used only if the construction of the final column provides some protection against easy breaking. The second type of tubing has—for inside diameters up to 0.50 mm–a uniform outside diameter of about $\frac{1}{16}$ inch (1.57 mm), i.e., a much thicker wall, which results in better mechanical strength.

3.2 COATING OF THE COLUMN TUBING

The column tubing must first be cleaned in order to eliminate any material that may have remained inside the tubing from the manufacturing process. It is then coated with a thin film of the stationary phase.

The coating techniques reported in the literature can be divided into two groups, the dynamic and the static methods. In both, the stationary phase is applied as a solution in an organic solvent. The solvent is then evaporated, and the dissolved stationary phase is thus deposited on the inside column wall. After that, certain conditioning is necessary before the column can be used for analysis.

3.21 Cleaning of the Column Tubing

Most of the metal tubing on the market which can be used for the preparation of open tubular columns has—as a result of the manufacturing process—some liquid adsorbed on the inside wall. The amount of this original "coating" can be fairly high. For example, Porcaro [222] measured 26.5 mg of material, found to be mainly polyisobutylene, in 200 ft of 0.50 mm ID copper tubing, which corresponds to a liquid-film thickness of approximately 0.32 micron.

Because foreign substances are almost invariably present in the tubing, a cleaning step prior to coating is a necessity. This cleaning procedure can be accomplished by forcing an organic solvent through

---

[31] From *Tubing for Gas Chromatography*, Bulletin No. 110, Superior Tube Co., Norristown, Pa.

the tubing in a manner similar to the dynamic coating procedure discussed below. The most convenient cleaning solvents are methylene chloride, chloroform, acetone, methanol, hexane, and diethyl ether. The flushing sometimes must be repeated several times with different solvents. For example, Hollis [138] recommends the following order of solvents: pentane, methylene chloride, acetone, diethyl ether, and the solvent of the stationary phase.

After washing, the tube is dried in a stream of the inert gas prior to coating.

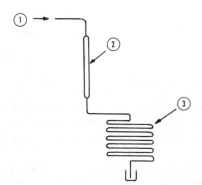

Figure 39. Equipment for coating open tubular columns by the dynamic method. *1* Inert gas, *2* reservoir for the coating solution, *3* column tubing.

3.22 Dynamic Coating Method

This method was used originally in the early work of Golay, but it was first described by Dijkstra and de Goey [68]. Figure 39 is a schematic diagram of the necessary equipment. In this method, the solution of the stationary phase is forced through the tubing with the aid of a dry inert gas such as argon, helium, or nitrogen. Thus, the inside wall of the tube is wetted by the solution. Subsequently the solvent is evaporated by blowing dry gas through the column for a few hours.

Two variations of the dynamic coating method are used, distinguished by whether the volume of the solution of the stationary phase is larger or smaller than the volume of the tubing itself. In both cases, it is important that the linear velocity of the coating solution be kept constant during its passage through the tubing and that a sudden change in velocity as the last of the solution leaves the column be avoided. Most authors advise using low linear velocities: according to Desty *et al.* [63], it should be always less than 10 cm/sec, and Kaiser

[160] used velocities of about 2 cm/sec. This requires fairly low flow rates (e.g., a linear velocity of 10 cm/sec corresponds to flow rates of 0.29 and 1.18 ml/min for columns of 0.25 and 0.55 mm ID, respectively), which are not easy to regulate accurately. Therefore, Kaiser [160] developed a special "microelectrolytic pump" for this purpose; however, in practice, the inlet pressure is usually regulated rather than the flow rate.

Since the tubing used for the preparation of open tubular columns usually has a small diameter, it is very important that the coating solution be free of solid particles, since they could easily plug the tubing. Therefore, the coating solution should be filtered immediately prior to use. In addition, if the end of the column tubing is submerged in a small vial filled with water or the solvent used (see Figure 39), continuous bubbling will indicate that the coating solution is moving unhindered through the tubing.

In both variations of the dynamic coating method, a proper balance of the concentration of the stationary phase in the solution and the travel time along the column will secure the appropriate liquid-film thickness. Generally, it can be stated [160] that above about 0.5 cm/sec the thickness of the coated film increases with the linear velocity and the concentration of the coating solution and also with the viscosity of the solvent, the other two variables being constant. The polarity of the solvent also influences the coating: according to Kaiser [160], if the concentration and the linear velocity are kept constant but two different solvents with the same viscosity are used, the polar solvent will result in a thicker coating. Change in the diameter of the column tubing has an inverse effect on the film thickness: increase of the diameter of the column—keeping the linear velocity and the concentration of the coating solution constant —will result in a decrease of the film thickness.

The advantages of the dynamic coating method are that it does not require any special equipment, that it can be carried out very easily, and that tubing coiled into the final form can be coated without difficulty. For these reasons, it is today the most widely used method. However, it has the disadvantage that film thickness values (and the values of the β term) are not readily available and that their measurement is difficult.

3.221   Volume of Coating Solution Larger than the Volume of Tubing.   In this first variation of the dynamic coating technique,

the volume of the coating solution is larger than (or equal to) the volume of the column tubing. Different workers differ somewhat in details of experimental technique: some push a volume larger than the volume of the tubing through it, while others start with a larger volume but disconnect the reservoir as soon as the first drop of solution appears at the column end (i.e., the column is completely filled with the solution).

Most workers use a 10% solution (or 10 g/100 ml) of the liquid phase and 15–40 psi (gauge) gas pressures upstream of the reservoir for pushing the solution through the tubing.

As mentioned above, the basic disadvantage of the dynamic method is that it does not permit an easy determination of the film thickness and the β term. The only real practical possibility is to weigh the column tubing before and after coating and thus determine the amount of coating in the column (M). In this way, the average liquid-film thickness (d_f) can be calculated from the density of the liquid phase (d_L), column radius (r), and length (L):

$$d_f = \frac{M}{2d_L r \pi L} \tag{55}$$

However, this method is not too accurate, since the weight of the column tubing is normally several orders of magnitude higher than the total weight of the coating. Another possibility might be to collect the leftover liquid phase solution and compare its volume with the starting volume; from the difference, the amount of liquid phase remaining in the column can be calculated.

If the average thickness of the liquid film is known, the β term can be calculated using Eq. (25):

$$\beta = \frac{r}{2d_f} \tag{25}$$

Sometimes values of the partition coefficient (K) at certain temperatures can be found in the literature. In such cases, by analyzing the same substance at the same temperature on the new column and determining the value of the partition ratio (k), the value of the β term can be calculated from Eq. (4):

$$K = \beta k \tag{4}$$

If the β value of an open tubular column is known, the β value of any other column with the same stationary phase can be obtained by analyzing a given substance on both columns at the same temperature; in this case

$$\beta_1 k_1 = \beta_2 k_2 \tag{56}$$

The values of k_1 and k_2 can be measured, and β_1 is known; thus, β_2 can be calculated.

3.222   Plug Method.

In the second variation of the dynamic coating method, the volume of the solution of the stationary phase is *less* than the volume of the column tubing, and the solution travels as a plug through the column at constant linear velocity, wetting the inside surface. For example, Scott and Hazeldean [249] filled 200 cm of 100-ft-long 0.5–1.0 mm ID columns (i.e., 6.7% of their total volume) and pushed this plug along the column at a linear velocity of 2–5 mm/sec. After the plug passed the column, the flow of inert gas was raised slowly to 1 ml/min and held for a time in order to evaporate the solvent from the coating.

The most detailed investigations of this method were carried out by Kaiser [160]. According to him, the linear velocity values should be about 2–5 cm/sec mainly because at the low values used by Scott and Hazeldean a very slight change in the velocity would result in a significant change in the film thickness, and since the exact regulation of such small flows is almost impossible, this fact would make the reproducible preparation of columns very difficult. As a conclusion of his investigations, Kaiser published some empirical equations for the necessary volume of coating solution (V_s) and the average film thickness (d_f) obtained with hexane as the solvent as functions of the column inside diameter (d) and length (L), the concentration of the coating solution (c), and the linear velocity of the solution in the tubing (u_s):

$$d_f = \frac{c}{100d}(0.265u_s + 0.25) \tag{57}$$

$$V_s = \frac{\pi}{c}Ld_f d = \pi L(0.265u_s + 0.25) \tag{58}$$

In these equations, the dimensions of the various parameters are as follows:

V_s:	mm^3	d:	mm
c:	vol-%	d_f:	μ
u_s:	cm/sec	L:	m

For example, if $c = 10\%$, $d = 0.25$ mm, $u_s = 2$ cm/sec, and $L = 50$ m, then $d_f = 0.31\ \mu$ and $V_s = 122$ mm³, which represents 5% of the total column volume. Kaiser suggests that in practice one should use twice the volume of V_s for coating.

The actual determination of d_f and β can be accomplished by methods similar to those described in chapter 3.221, or else by a relatively easy method described by Kaiser [160]. For this purpose a special apparatus as shown in Figure 40 is constructed. The crucial

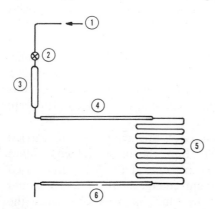

Figure 40. Apparatus for coating open tubular columns by the plug method. *1* Inert gas, *2* needle valve for flow regulation, *3* reservoir for the coating solution, *4* graduated transparent capillary tube, *5* column tubing, *6* graduated transparent capillary tube.

parts of this apparatus are the capillary tubes *4* and *6*. The internal diameters of these tubes should be identical or close to each other, and both should be calibrated exactly to volume. The coating solution is introduced into reservoir *3* and then pushed into the first capillary tube *4*, where its volume is exactly measured (V_{s1}). Now, the solution is pushed through the column tubing *5* into the second capillary tube *6*, where its volume is again measured (V_{s2}). The difference $V_{s1} - V_{s2}$ will give the volume of the solution remained in the column tubing. The average liquid-film thickness (d_f) can be calculated if the column length (L), internal diameter (d), and concentration (c, vol-$\%$) of the coating solution are known:

$$d_f = \frac{(V_{s1} - V_{s2})c}{100\pi L d} \tag{59}$$

In the construction of this apparatus, it is very important that the connections of the column tubing to the graduated capillary tubes *4* and *6* be kept as short as possible since the connecting tubes will obviously also be coated, with a resultant loss of coating solution.

3.23 Static Coating Method

This coating method, which was first developed by Golay [110] and used in his original systematic work, is based on the following principles. The column is first completely filled with a solution of the liquid phase (using a system similar to that shown in Figure 39). Then one end of the tubing is closed, following which the tubing is slowly drawn through an oven, open end first, in order to evaporate the solvent from the column.

The static coating method has the disadvantages that it requires relatively complicated equipment, that it is itself more complicated, and that the column tubing can be formed into its final shape only *after* coating. These are the main reasons why it was not used in general practice. Recently, however, Horváth [141] extended its application to the preparation of open tubular columns having a thin layer of porous support or adsorbent deposited on the inside wall (see chapters 3.42 and 3.5).

The outstanding advantage of the static coating method is that —since the solution completely fills the tube—the value of β can be calculated directly from the volume concentration of the liquid phase solution. For example, if a 1% by volume solution is used for coating a 0.25 mm ID ($r = 125\ \mu$) column,

$$\beta = \frac{V_G}{V_L} = \frac{99}{1} = 99 \tag{60}$$

and, using Eq. (25),

$$d_f = \frac{r}{2\beta} = \frac{125}{198} = 0.63\ \mu \tag{61}$$

It is evident from Eqs. (60) and (61) that the value of β is independent on the diameter of the column tubing and is governed solely by the concentration of the coating solution. If solutions with the same concentration are used, the thickness of the stationary film coated on the inside wall of the tubing is directly proportional to the tube radius. Since generally, a film thickness of 0.5–1 μ is desired in practical columns, coating solutions with a concentration of 2% or less have to be used.

The comparison of the values calculated from Eqs. (60) and (61) with those obtained from Eq. (57) shows that in order to achieve similar film thicknesses, a ten- to twentyfold increase in the concentration of the coating solution is required for the dynamic (plug) coating method as compared to the static method.

3.24   Elimination of the Secondary Adsorption Effects of Column Tubing

It is generally known that practically all support materials (except Teflon) show some polarity which might result in secondary adsorption effects and peak tailing (distortion of the ideal Gaussian peak). Therefore, for diatomaceous-earth-type supports, the method of silanization[32] is used to block the active sites of the support particles.

The situation is similar for work with open tubular columns, where interaction between the sample molecules and the inside wall of the tubing can result in—sometimes severe—tailing. Kiselev [171] pointed out that even glass tubing can show such effects and suggested silanization of the glass column tubing itself as a precautionary measure.

Secondary adsorption effects are naturally strongest toward polar sample molecules. Due to the tendency of the polar surface to attract strongly polar materials, it will be only poorly wetted by a nonpolar stationary phase. As a result, such a phase will more readily bleed off at elevated temperatures, resulting in lifetimes shorter than expected.

In 1961, Averill [10] proposed the use of small amounts of surface-active materials as additives to the stationary phase to eliminate these undesired effects of the column wall. These surfactants are substances with strongly polar groups at one end and a long hydrocarbon chain at the other. A typical compound is Atpet 80, which is chemically sorbitan monooleate:

$$CH_3(CH_2)_7CH{=}CH(CH_2)_7COOC_6H_8(OH)_5$$

Here, the five free hydroxyl groups at one end are the strongly polar groups which will be permanently adsorbed by the column wall. A similar substance is Span 80.

The substances which can be used as additives can be divided into three groups. The first group consists of nonionic substances. These are usually polyglycols such as sorbitol which have been partially esterified with a long-chain fatty acid. Atpet 80 and Span 80 are typical examples of this group. Such additives can generally be used for both nonpolar and polar liquid phases and samples.

The two other groups consist of additives of anionic or cationic character. The structure of these substances is essentially similar to

---

[32] See, e.g., D. M. Ottenstein, *J. Gas Chromatog.* **1**(4):11–23 (1963).

that shown above, except that the active groups, respectively, are of carboxylic character or contain nitrogen atoms. Additives of these groups can be used successfully when analyzing substances of strong anionic (acidic) or cationic character.

According to Averill, these additives are permanently adsorbed on the inside wall, thus inactivating its active sites. At the same time, the long chain will remain loose, resulting in a velvetlike structure as idealized in Figure 41. This structure will act as a "glue" to span the interface between the stationary phase and the surface of the support: the phase will wet only this "velvet," and the sample molecules will not reach the active sites of the wall.

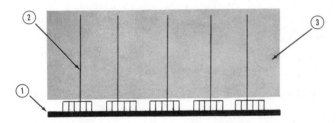

Figure 41. Idealized schematic of the coated column inside wall using surface-active additives to the liquid phase. *1* Column wall, *2* additive molecule, *3* liquid coating.

The effect of such an additive is illustrated by a comparison of Figure 42 with Figure 43.[33] Both columns were coated with squalane stationary phase from a 10% solution; however, in the second case, Alkaterge T and Span 80 surfactants (each in 0.1% concentration) were added to the solution. As a result, the tailing of the ethanol peak is dramatically reduced.

These surfactants start to bleed off the liquid phase around 170°C, and when the column is operated above this temperature for an appreciable period, they will be completely eliminated. Experience, however, has shown that even in the case of columns which are generally used above 170°C, the addition of additives to the coating solution has definite advantages. This fact might be explained by the hypothesis of Farré–Rius *et al.* [92] that they reduce the surface

---

[33] In both cases, the so-called polarity mixture was used as sample. It is discussed in more detail in chapter 3.3.

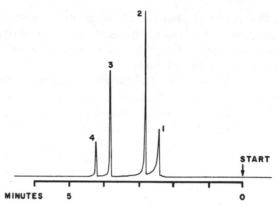

Figure 42. Analysis of the polarity mixture. I [77]. Column: 150 ft × 0.25 mm ID open tubular coated with squalane liquid phase. Column temperature: 50°C. Peaks: *1* ethanol, *2* methyl ethyl ketone, *3* benzene, *4* cyclohexane.

tension of the liquid phase and thus facilitate the spreading of the liquid.

In most cases, a concentration of 0.1–0.2% of the additive in the coating solution (as compared to a concentration of 10% of the stationary phase itself) is adequate. However, for the analysis of highly polar samples, this amount of additive might not be sufficient and surfactants have to be used in fairly high concentrations. The free fatty acids and phenol homologs belong to this sample category. For either, no satisfactory analysis is possible on open tubular columns without the use of large amounts of additives.

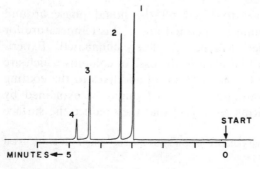

Figure 43. Analysis of the polarity mixture. II [77]. Column: 150 ft × 0.25 mm ID open tubular coated with squalane liquid phase, with Alkaterge T and Span 80 additives. Column temperature: 50°C. Peaks: *1* ethanol, *2* methyl ethyl ketone, *3* benzene, *4* cyclohexane.

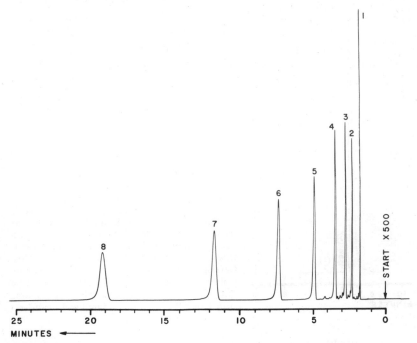

Figure 44. Analysis of free fatty acids [11]. Column: 200 ft × 0.50 mm ID open tubular coated with Trimer Acid liquid phase, with dinonylnaphthalenedisulfonic acid additive (10/0.4, w/w). Carrier gas (He) flow rate at outlet: 15 ml/min. Column temperature: 140°C. Peaks: *1* acetone (solvent), *2* acetic acid, *3* propionic acid, *4* butyric acid, *5* valeric acid, *6* caproic acid, *7* heptylic acid, *8* caprylic acid.

Figure 44 shows the analysis of C_2–C_8 free fatty acids in acetone solution [11]. In this case, Trimer Acid[34] was used as liquid phase, with dinonylnaphthalenedisulfonic acid—an anionic substance—as additive. The concentration of the additive in the coating solution was 0.4%, as compared to 0.1–0.2% used in the previous example. The practical application of such a column is mainly in the analysis of very diluted water solutions or of technical acids with a not too wide range. For example, Figure 45 shows the analysis of a technical acrylic acid with acetic and propionic acid impurities.[35] The amount of additive is now even higher than in the previous case.

[34] Trimer Acid is a C_{54} tribasic acid with about 10% of C_{36} dibasic acid content; a commercial product of Emery Industries, Cincinnati, Ohio.

[35] G. Oesterhelt (Bodenseewerk Perkin–Elmer & Co., Überlingen, Germany), unpublished result.

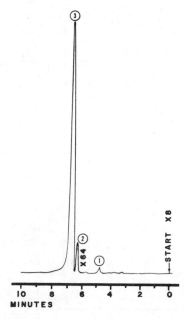

Figure 45. Analysis of a technical acrylic acid.[35] Column: 50 meter × 0.25 mm ID open tubular coated with Trimer Acid liquid phase, with dinonylnaphthalenedisulfonic acid additive (10/1, w/w). Sample volume: 0.4 μl (liquid), split 1/100. Carrier gas (He) flow rate at outlet: 0.8 ml/min. Column temperature: 130°C. Peaks: 1 acetic acid, 2 propionic acid, 3 acrylic acid. Flame ionization detector with 2.5 millivolt recorder.

As already mentioned, another group of substances whose satisfactory analysis with open tubular columns is made possible only by the use of additives are the phenol homologs. Figures 46 [12] and 47[36] give an example on the results which may be obtained. Both columns were coated with a 10 g/100 ml solution, and the relative amounts of stationary phase and additive are given in the figure captions.

At the moment, there is practically only one significant organic substance group which has not been analyzed on open tubular columns with satisfactory results. This group consists of the various steroids, whose gas chromatographic analysis became one of the most important new achievements in biochemistry. Although Lipsky and Landowne [180] demonstrated two chromatograms obtained on open tubular columns with neopentyl succinate liquid phase, these chromatograms are far poorer than those obtained routinely on packed columns. The principal reason for the poor performance of open tubular columns in this instance may be that at the high analysis

---

[36] Chromatogram reproduced from Column Sheet GC–110 of Bodenseewerk Perkin–Elmer & Co., Überlingen, Germany.

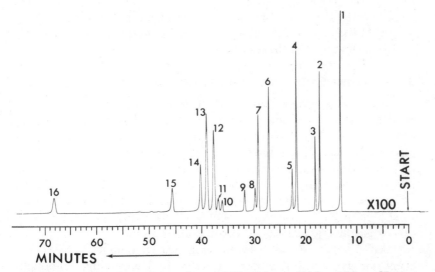

Figure 46. Analysis of phenol homologs. I [12]. Column: 150 ft × 0.25 mm ID open tubular coated with didecyl phthalate liquid phase, with Atpet 80 additive (85/15, w/w). Column temperature: 120°C. Peaks: *1* phenol, *2* o-cresol, *3* 2,6-xylenol, *4* p-cresol, *5* m-cresol, *6* o-ethylphenol, *7* 2,4-xylenol, *8* 2,5-xylenol, *9* 2,4,6-trimethylphenol, *10* 2,3-xylenol, *11* 2,3,6-trimethylphenol, *12* p-ethylphenol, *13* m-ethylphenol, *14* 3,5-xylenol, *15* 3,4-xylenol, *16* 3-methyl-5-ethylphenol.

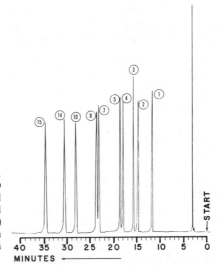

Figure 47. Analysis of phenol homologs. II.[36] Column: 50 meter × 0.25 mm ID open tubular coated with didecyl phthalate liquid phase, with Trimer Acid additive (90/10, w/w). Carrier gas (He) flow rate at outlet: 1.5 ml/min. Column temperature: 130°C. Peaks: as given in Figure 46.

temperature (220–230°C) the active sites of the column wall are no longer blocked. To date, however, no high-temperature surfactant has been described in the literature.

3.25 Column Conditioning and Storage

After the coating is finished, the column is first dried in a stream of inert gas for a few hours. After that, the column is placed into an oven (or into the gas chromatograph itself) and the temperature is raised at a very slow rate (1–2°C/min) to a level about 25°C lower than the so-called recommended maximum temperature of the particular stationary phase. The column is kept at this temperature for a few hours (preferably overnight), and is then cooled down slowly to room temperature. During this entire process, a slow carrier gas flow is maintained through the column.

As a result of this conditioning, the remaining solvent will evaporate and possibly present unstable or low-boiling compounds (contaminants) in the stationary phase will also pass from the column. The conditioning can be monitored with the recorder; a stable baseline reflects the end of the emission.

Upon completion of the conditioning, the column should be stored by sealing the two ends in order to avoid contact of the stationary phase film with an oxidizing atmosphere. In this way, deterioration of the column can generally be avoided. There are only a few reports of changes in the stationary phase during storage, mainly with polyester phases, where impurity traces remaining in the column might catalyze decomposition, or with stationary phases which are solid at room temperature (e.g., 7,8-benzoquinoline). In the latter case, the possibility that recrystallization may result in globules rather than the original uniform coating film has been suggested [232].

3.26 Recoating of Open Tubular Columns

After a certain time, any gas chromatographic partition column will lose part of its stationary phase content due to "bleeding," i.e., slow evaporation into the continuously moving carrier gas stream. The rate of loss is accelerated by operating the column at elevated temperatures, because the vapor pressure of the stationary phase (and therefore the bleeding) will increase exponentially with temperature. Sometimes certain sample components may catalyze the chemical decomposition of the stationary phase; also, if for any reason oxygen (air) is introduced at elevated temperatures, it will result in oxidation of the stationary phase. As a result, all columns must be regenerated periodically.

The regeneration of open tubular columns consists of two steps. First, the remaining stationary phase has to be eliminated; then, the column has to be recoated.

The elimination of the rest of the stationary phase is usually performed by repetitive washings with various solvents. Sometimes an acid solution is necessary to destroy the scales formed by the remaining stationary phase. For example, Averill [8] recommends the consecutive use of methylene chloride, water, methanol, water, 20 % nitric acid, water, methanol, and methylene chloride. Generally, the last solvent should be that which is used in the preparation of the new coating solution.

After washing, before coating, the column is dried in a stream of the inert gas.

Washing can be carried out in a system similar to that shown in Figure 39. Almost all liquid phases used in open tubular columns can be washed out. The only exceptions might be the various silicone gum rubber types, which sometimes repolymerize at high operating temperatures, actually resulting in a higher-molecular-weight material than that originally used for coating.

Generally, the dynamic method is used for column recoating, particularly if the column was originally wound into a certain shape, since the static coating method would require rewinding of the tubing.

3.3 TESTING OF OPEN TUBULAR COLUMNS

There are many possibilities for testing open tubular columns. One can use a single component and calculate the theoretical plate number, the effective plate number, or only t_R/w_h. In this case, a peak with $k \geq 3$ should be selected. Another possibility is to use a typical sample mixture and calculate plate number and resolution. This is particularly recommended for specific columns where one particular separation (e.g., stearate/oleate, m/p-xylene) is critical.

The use of a typical test mixture is also recommended for periodic checking of column performance during use. Naturally, for each test mixture, the analytical conditions must be fixed.

3.31   The Use of the "Polarity Mixture" for Testing

A special method developed originally by Averill [10] is recommended for practical column testing [77]. In this method, a so-called *polarity mixture*, the composition of which is given in Table XIII, is used as sample. As can be seen, the boiling points of all four components are close to each other; their polarity, however, differs greatly: ethanol has the highest polarity, followed by methyl ethyl ketone and benzene, while cyclohexane is not polar at all. The reason for the selection of very different concentrations was to facilitate the identification of the individual peaks.

TABLE XIII
Composition of the "Polarity Mixture"
Used for Column Testing [77]

Component	Boiling point, °C	Parts by volume
Ethanol	78.5	40
Methyl ethyl ketone	79.6	20
Cyclohexane	81.4	5
Benzene	80.1	10

When this polarity mixture is analyzed on an open tubular column, the chromatogram gives much information regarding the polarity of the stationary phase, the secondary adsorption effects of the inside column wall, the separating ability of the column, the order of emergence of substances of different classes but similar boiling points, and the adequacy of the liquid-film thickness.

The polarity of the stationary phase is indicated by the sequence of the peaks. On a nonpolar column (e.g., squalane), the order of elution is ethanol, methyl ethyl ketone, benzene, and cyclohexane, while on a slightly polar column (e.g., phenyl silicone oil), cyclohexane will come before the benzene peak. On the other hand, on polar liquid phases (e.g., Carbowax 1540 polyethylene glycol), cyclohexane will be the first peak. Figures 42, 43, 48, and 49 demonstrate this effect on three different columns coated with the stationary phases mentioned.[37]

---

[37] In the columns used for the analysis shown in Figures 43 and 48, Atpet 80 was added to the liquid phase (see chapter 3.24).

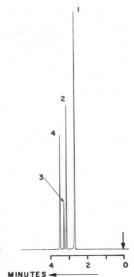

Figure 48. Analysis of the polarity mixture. III [77]. Column: 150 ft × 0.25 mm ID open tubular coated with DC-550 phenyl silicone oil liquid phase (with Atpet 80 additive). Column temperature: 100°C. Peaks: *1* ethanol, *2* methyl ethyl ketone, *3* cyclohexane, *4* benzene.

The second piece of information to be gained from a test analysis of the polarity mixture concerns the secondary adsorption of the effect of the inside wall of the column tubing. If the ethanol and methyl ethyl ketone peaks tail, it is mainly due to this effect. Such tailing can be seen, e.g., in Figure 42. Chapter 3.24 discussed how this tailing can be eliminated.

The degree of resolution between the cyclohexane and benzene peaks is an indication for the separation between paraffins or cyclo-paraffins and aromatics; similarly, the resolution of the methyl

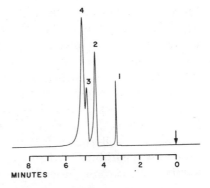

Figure 49. Analysis of the polarity mixture. IV [77]. Column: 200 ft × 0.50 mm ID open tubular coated with Carbowax 1540 liquid phase. Column temperature: 50°C. Peaks: *1* cyclohexane, *2* methyl ethyl ketone, *3* benzene, *4* ethanol.

ethyl ketone and alcohol peaks is an indication of the behavior of the column for mixtures of ketones and alcohols in general: some stationary phases, e.g., the polyglycols (see Figure 49), retain ketones much longer than alcohols with similar boiling points.

Finally, tailing of the cyclohexane or methyl ethyl ketone peak could be an indication of either dead volume in the system or of a too thick or too viscous liquid phase. This can be seen, for example, in Figure 49, where the methyl ethyl ketone peak shows definite tailing; this column actually had too much liquid phase.

3.32   Determination of the "Air Peak" Time

If chromatograms are to be compared with each other, relative retention times evaluated, or various characteristic parameters (e.g., the partition ratio) calculated, the passage time of the carrier gas front through the column must be known. In practice, the retention time of an inert component (usually air) which is insoluble in the stationary phase and, therefore, is not retarded upon passage through the chromatographic column is taken as this value (t_M).

When thermal conductivity detectors or gas density balances are used, the retention time of air is readily available. However, ionization detectors generally do not respond to air and therefore, in this case, evaluation of the chromatogram in this manner is not possible.

In such cases, methane is usually substituted for air and the retention time of methane is taken as t_M. For this purpose, methane has to be injected into the gas chromatograph separately but under identical conditions (temperature, flow rate). In most cases, this substitution is satisfactory, but it must be remembered that *methane is soluble in almost all stationary phases*, although—particularly at higher temperatures—it has only a very small k value, i.e., its retention time differs only very slightly from that of air. However, sometimes— mainly with liquid phases used only at room or slightly elevated temperatures for the analysis of hydrocarbons—a definite error is introduced into all calculations.

In the practical samples analyzed in gas chromatography, members of various homologous series are usually present. With help of these peaks the position of the "air peak" can be calculated, utilizing the well-known fact that a straight-line relation exists between the logarithms of the adjusted retention times (log t_R') and the number of carbon atoms in a homologous series. Since the determination of the "air peak" time (t_M) is a common problem in gas chromatography with open tubular columns and ionization detectors, the methods used for this calculation are given in the Supplement.

3.4 INCREASE OF THE INSIDE SURFACE AREA OF THE COLUMN TUBING

As discussed in detail in the second part of this book, the shortcomings of open tubular columns are their very small sample capacity and the relatively large β term. Both are related to the fact that the amount of liquid phase in the column is very small and cannot be increased without resulting in much too thick a film. Such thick films are not stable; moreover, the liquid diffusion term could no longer be neglected. Columns of larger diameter of course have a larger sample capacity; but the efficiency of such columns is lower than that of columns of smaller diameter, and the value of the β term is increased even more.

Another way to increase the sample capacity—which, at the same time, would decrease the value of the β term—is to increase the inside surface of the column without increasing its diameter. This method was proposed by Golay as early as 1958 [110], and in 1960 he reported on some preliminary results [111].

There are two ways to increase the inside surface area of the column tubing. In the first case, the inside surface of the tube is treated chemically in order to produce a rough surface, while in the second case, a porous support layer is deposited on the inside wall. Recently some encouraging reports have been published on both methods. Both techniques can also be used for the preparation of open tubular *adsorption* columns, and thus investigations in the two areas are somewhat interlinked and most workers who prepared columns with these techniques try to use them both coated and uncoated, i.e., both as partition and adsorption columns.

Although all these developments are at the moment still in the experimental stage, they could very well represent an important new concept in the construction of open tubular columns. Therefore, these new results are discussed here in more detail than their present importance would seem to justify.

3.41 Chemical Treatment of the Inside Wall of the Tubing

The first successful method for preparing a rough inside surface in *glass* tubing was described by Mohnke and Saffert [200], but the columns produced in this way were primarily used in gas solid adsorption chromatography (see chapter 3.5). According to a later publication, however, such columns were also coated with a thin film of liquid phase [177], but detailed data on the performance of these columns are not available.

In the original method used by Mohnke and Saffert, a glass tube of 0.27 mm internal diameter was filled with a 17% aqueous ammonia solution, sealed at both ends, and then heated to 170°C for 70 hours. After this treatment, the tube was opened and the ammonia solution removed by means of compressed air and the temperature of the tube raised slowly to 190°C, until it was dry and annealed. During this period, a constant air stream was maintained through the tubing. With this treatment, a silicon dioxide layer of about 20μ thickness was formed on the inner wall of the tubing.

According to the second publication [177], the method was slightly modified. Now, a 12% aqueous ammonia solution was used, the sealed tube was heated to 170–180°C for 30 hours, and the drying period took 36 hours. In this way, a SiO_2 layer of approximately 5μ thickness was formed.

More detailed investigations on treated glass tubing were reported recently by Bruner and Cartoni [30]. In their method, a 20% aqueous NaOH solution is pushed through the glass tubing (0.44 mm ID) from a reservoir for 6 hours at a flow rate of 10 ml/hour while the temperature of the tube is maintained at 100°C. The tubing is then washed with distilled water until no alkaline reaction is observed in the effluent, followed by washings with absolute ethanol and ether, and finally the tubing is dried in a flow of nitrogen. As a result of this treatment, a porous layer is formed on the inside wall. Tubing treated in this way was coated with solutions of different concentrations of squalane liquid phase, so that various average liquid-film thicknesses ranging from 0.01μ to 0.18μ were obtained. Measurements showed that the value of k was not directly proportional to the film thickness, which indicates that the retardation of the simple components in the column was influenced not only by the partitioning process but also by certain adsorption effects of the porous glass wall; this is also indicated by the fact that an identical column with zero coating actually showed a larger value of k than columns with thin film coatings. The β values varied from 1471 ($d_f = 0.01 \mu$) down to 92 ($d_f = 0.18 \mu$). The actual volume of liquid phase used in treated columns was 7.1 times that required to maintain the same average liquid-film thickness in a 0.44 mm ID open tubular column without any wall treatment. Of course, this means that the sample capacity is also increased by this factor.

Wall-treated columns coated with squalane liquid phase can be used successfully for the separation of cyclohexane (C_6H_{12}) and deuterocyclohexane (C_6D_{12}). At 40°C and a carrier gas (N_2) outlet flow of 1.25 ml/min, with a 38-meter-long column with an average liquid-film thickness of 0.12μ and a β value of 131, the relative retention (α) of the two components is 1.07 and a baseline separation ($R = 1.45$) can be obtained.

So far we have spoken only of the treatment of the inside wall of glass tubing; however, Jentzsch and Hövermann [154] have reported the successful treatment of *copper* columns. In their method, the copper tube (100 meter × 1 mm ID) is flushed with 40% aqueous nitric acid solution. After cleaning with distilled water, the column inside wall is oxidized by means of dry oxygen gas at 200–250°C for 5–7 hours. In this way, a spongelike inside surface is obtained, which can be coated with the solution of any liquid phase, using the methods described in chapters 3.221 and 3.23. Experience has shown that coating by the plug method does not result in satisfactory columns.

Investigations with both untreated and wall-treated copper columns coated with squalane liquid phase showed that the sample capacity of the latter is four times that of the untreated columns. ("Sample capacity" here means the sample volume up to which only less than 10% reduction in the theoretical plate number can be observed.) Further experience with binary sample mixtures has shown that, e.g., in case of the 2- and 3-methylpentane pair, a 16 μl (liquid) sample injected into the wall-treated column (carrier gas flow rate: 33 ml/min) results in the same resolution as a 0.5 μl sample in the untreated column.

3.42   Deposition of a Porous Support Layer on the Inside Wall of the Tubing

While in the previous methods, the inside wall of the tubing is modified by a chemical treatment, here a porous inert support layer is deposited on the inside wall. Golay [111], in 1960, reported on some promising preliminary investigations carried out at Perkin–Elmer by depositing a layer of colloidal clay on the wall of the tube, firing the clay *in situ*, and impregnating it with a stationary phase using the conventional coating methods. His report, however, was

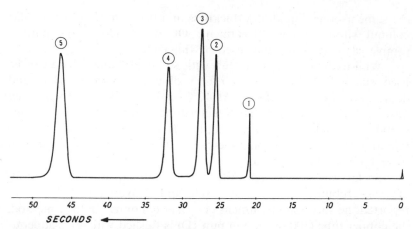

Figure 50. Analysis of a hydrocarbon mixture. I [124]. Column: 30 meter × 0.283 mm ID open tubular coated with squalane liquid phase. Carrier gas (H$_2$) inlet pressure, flow rate at outlet, and average linear velocity: 2 atm (gauge), 11.8 ml/min, 144 cm/sec. Column temperature: 80°C. Peaks: *1* methane, *2* n-pentane, *3* 2,2-dimethylbutane, *4* n-hexane, *5* n-heptane.

not followed immediately by other researchers. Finally, in early 1963, Halász and Horváth published their results [123, 124, 141] on a new method for depositing fine solid particles on the inside wall of the column tubing. If the solid was an adsorbent, the columns prepared in this way could be used in gas adsorption chromatography (see the next chapter); on the other hand, by using finely powdered support

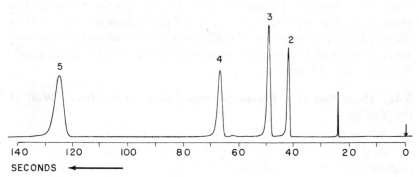

Figure 51. Analysis of a hydrocarbon mixture. II [124]. Column: 30 meter × 0.508 mm ID open tubular prepared with a thin support layer on the inside wall and coated with squalane liquid phase. Carrier gas (H$_2$) inlet pressure, flow rate at outlet, and average linear velocity: 1 atm (gauge), 23.8 ml/min, 128 cm/sec. Column temperature: 80°C. Peaks as in Figure 50.

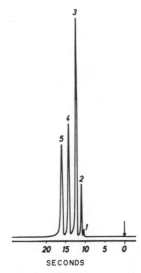

Figure 52. Analysis of light hydrocarbons [141].
Column: as in Figure 51. Carrier gas (H_2) inlet pressure
and flow rate at outlet: 1.5 atm (gauge), 77 ml/min.
Column temperature: 62°C. Peaks: *1* methane, *2*
ethane + ethylene, *3* propane + propylene, *4* isobutane,
5 n-butane.

material and coating it with a liquid phase, efficient partition columns
could be obtained.

The method described by Halász and Horváth is based essentially
on the static coating technique described in chapter 3.23. The
stationary phase is dissolved in an appropriate solvent, the solution
is mixed with the fine diatomaceous earth support, and the solvent is
evaporated following the technique employed in preparing column
packing material. The coated support so obtained is then mixed in
a high-speed mixer with a high-density organic liquid (e.g., a mixture
of methylene bromide and carbon tetrachloride) until a stable
emulsion is formed. Then the column tubing is filled with this
emulsion and the solvent is evaporated. When a copper column of
0.5 mm internal diameter is coated in this way, the inner surface area
of the tube is increased by a factor of 130. At the same time, the value
of the β term for this column was found to be only 35.5. The HETP
values measured were very close to those obtained for a "classical"
wall-coated 0.5 mm ID open tubular column ($d_f = 0.6\,\mu$), while the
values of k were larger by a factor of about 3.4. Figures 50 and 51
compare the chromatograms obtained on the "classical" open
tubular column and the column prepared with a coated thin support
layer, and Figure 52 shows the fast analysis of some light hydrocarbons on the same column.

The performance of open tubular columns prepared with a thin support layer on the inside wall and coated with an appropriate liquid phase was also investigated recently by Norem.[38] In his work, finely powdered diatomaceous earth support and squalane liquid phase were used, and the column (ID = 0.5 mm) was prepared in a manner identical to that of Halász and Horváth. Some characteristic values of the column prepared by Norem are listed below:

Column internal diameter	0.5	mm
Geometric surface area of the column	15.7	cm^2/meter length
Surface area of the porous support	1800	cm^2/meter length
Average thickness of the porous support layer	60	μ
Average liquid film thickness	0.027	μ
Partition ratio for n-hexane at 50°C	5.7	
Value of β	27.9	

In this column, 42.2% of the cross section of the original tube was occupied by the porous thin layer.

This column may be compared with a "classical" open tubular column of the same internal diameter. The same volume of liquid phase in the latter would give an average liquid-film thickness of 3.09 μ, which is much too large; on the other hand, if the average thickness of the liquid film is kept constant (i.e., 0.027 μ), its volume in the "classical" column would be only 0.86% of that in the support-coated column. In this case, the value of the β term would be 4630 and the k for n-hexane at 50°C 0.034.

These results demonstrate the vast potentialities of open tubular columns in which a thin layer of a porous support is deposited on the inside wall in order to increase the surface area available for coating purposes.

Finally, it should be mentioned that for such columns the classical van Deemter—Golay equation [Eqs. (19)–(22)] can be used only with certain modifications. Recently Golay, in a short paper [113], discussed this question and showed that, e.g., in the case of a column in which the cross section of the porous layer equals 20% of the unobstructed cross section, the dynamic diffusion term [i.e., $C_G + C_L$ in Eq. (19)] will be 15% larger (for $k = 3$) than it would be in a "classical" tubular column.

---

[38] S. D. Norem (Perkin–Elmer Corp., Norwalk, Conn.), unpublished results.

3.5 ADSORPTION-TYPE OPEN TUBULAR COLUMNS

Recently, a number of publications have attempted to revive gas adsorption chromatography, and some of them have discussed different ways for preparing open tubular adsorption columns for the analysis of both inorganic and organic samples. It can be expected that in the near future more and more workers will deal with the performance of such columns. At the moment, their application is restricted to not too polar samples and substances with relatively low boiling point, since most of the adsorbents used desorb strongly polar molecules only at much too high temperatures, and even there, with significant tailing. Also some catalytic effects resulting in decomposition of olefins have been reported. However, there are indications that proper modification of the adsorbents could solve this problem; at least some preliminary results seem to be promising.

Below, the various methods for the preparation of adsorption-type open tubular columns that have so far been described in the literature are discussed, and applications of the different columns are illustrated. Finally, the necessary modifications in the general gas chromatographic theory are briefly outlined.

3.51 Plating Methods

These methods were described by Zlatkis and Walker [285]. Copper columns of 250 ft × 0.75 mm ID were generally used, and the tubing was first cleaned of any residual drawing oil. Next, an excess of plating solution was forced through the column at 10 psi (gauge) pressure. When the whole solution had passed the column, it was purged with a large amount of distilled water and acetone.

Silver plating was performed by using an aqueous solution of silver cyanide complex. First, a 0.1 molar silver nitrate solution was prepared to which excess sodium cyanide was added in order to dissolve the silver cyanide precipitate. Gold, platinum, and mercury plating solutions were made using 15% aqueous solutions of auric chloride, platinic tetrachloride, and mercuric nitrate, respectively.

Partial separations of various hydrocarbons have been reported with these columns, the most notable being the separation of *cis*- and *trans*-decaline at room temperature in about three minutes. All the chromatograms show definite tailing, however, and therefore an attempt has been made to modify the adsorbent surface by a thin liquid coating in a way similar to that in which Eggertsen *et al.* in

1956 modified some adsorbents by the addition of 1–1.5% liquid phase.[39] With such modified columns, reduction of peak tailing was observed.

3.52   Adsorption Columns with Chemically Modified Inside Wall

These columns are prepared by building up an adsorbent layer on the inside column wall by means of some chemical treatment of the tubing.

Preparation of glass columns with such a porous adsorbent layer was described first by Mohnke and Saffert [200]; their method is discussed in chapter 3.41. In addition to the utilization of this porous layer as support for a liquid phase, it was also used as an adsorbent in itself, for the separation of hydrogen isotopes and their nuclear spin isomers, at the temperature of liquid nitrogen (77.6°K). Figure 53 shows the chromatogram.

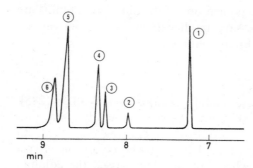

Figure 53. Separation of hydrogen isotopes and their nuclear spin isomers [200]. Column: 80 meter × 0.27 mm ID glass open tubular, with a 20 μ SiO$_2$ adsorbent layer on the inside wall. Carrier gas (Ne) flow rate at outlet: 2 ml/min. Sample size: 1.5 μl gas mixture. Column temperature: 77.6°K. Peaks: 1 He, 2 p-H$_2$, 3 o-H$_2$, 4 HD, 5 o-D$_2$, 6 p-D$_2$.

Bruner and Cartoni [30] also described modification of the inside glass surface by chemical treatment (see chapter 3.41). Although their columns were primarily used by coating them with a stationary phase film, the possibility of their application as adsorption columns —without any coating—was also demonstrated.

Both above-mentioned methods use an alkaline treatment for the formation of a porous adsorbent layer. On the other hand, Kiselev [171] utilized an acidic treatment for the same purpose. In his work, a 10 meter × 0.5 mm ID glass tube was etched with 0.1 normal hydrochloric acid at 25°C for 5 min and then washed with distilled water. This treatment resulted in a porous adsorbent layer

[39] F. T. Eggertsen, H. S. Knight, and S. Groennings, *Anal. Chem.* **28**:303 (1956).

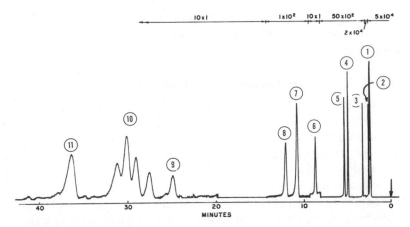

Figure 54. Analysis of a natural gas sample [217]. Column: 50 ft × 0.50 mm ID aluminum open tubular, with an activated alumina layer on the inside wall. Carrier gas (CO_2) flow rate at outlet: 2 ml/min. Sample volume: 5 ml gas, split 1/141. Column temperature: 100°C. Peaks: *1* methane, *2* ethane, *3* propane, *4* isobutane, *5* *n*-butane, *6* neopentane, *7* isopentane, *8* *n*-pentane, *9* cyclopentane, *10* isomeric hexanes, *11* *n*-hexane.

of about 100 μ thickness. Such columns were successfully used for the separation of C_1–C_4 hydrocarbons.

Modification of the inside surface of 0.50–0.625 mm ID *aluminum* tubes was also reported by Petitjean and Leftault [217], and today such columns are a commercial product of the Aluminum Corporation of America. By an undisclosed method, an about 5-μ-thick active alumina layer is formed on the inside wall of the column tubing. The surface area of this oxide film is about 4.6 m^2 per meter of column length. Successful application of such columns for the analysis of saturated hydrocarbons has been demonstrated; Figure 54, for example, shows the analysis of a natural gas sample at 100°C. Programming of the column temperature from 50 to 225°C helped in sharpening the peaks corresponding to the C_6 hydrocarbons and allowed the detection of C_7–C_9 hydrocarbons, which are permanently adsorbed by the column at 100°C.

The basic problem of these alumina open tubular adsorption columns is that they have a significant activity toward olefins, which in small concentrations will not pass the column at all. Petitjean and Leftault therefore proposed the modification of the active alumina adsorbent layer by a thin film of a high-boiling liquid. Further problems concern the abnormally large increases in retention time with increasing molecular weight and significant peak tailing for polar samples.

3.53 Open Tubular Columns Coated with an Adsorptive Layer

The most promising adsorption-type open tubular columns have been prepared by depositing an adsorbent layer on the inside surface of the column tubing. Except for Zlatkis and Walker [285], who prepared columns coated with potassium dichromate from its saturated aqueous solution, all the other researchers have utilized the colloidal solutions of finely powdered solid adsorbents.

Schwartz *et al.* [242] prepared columns with a porous silica layer using both plastic (Nylon and Delrin) and metal (copper and stainless steel) tubes of 0.50–0.85 mm internal diameter. The coating solution consisted of colloidal silica sol such as Nalcoag 1022, which is a 22% sol in a water–isopropanol mixture, or Cab-O-Sil suspended in water, cyclohexane, or benzene. Columns of various lengths were successfully used for the separation of C_1–C_7 saturated hydrocarbons

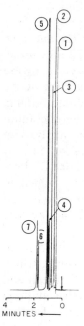

Figure 55. Analysis of light hydrocarbons.[40]   Column: 100 ft × 0.50 mm ID open tubular coated with a thin silica gel layer. Sample volume: 20 μl vapor. Column temperature: 150°C. Peaks: *1* ethane, *2* propane, *3* propylene, *4* isobutane, *5* *n*-butane, *6* isopentane, *7* *n*-pentane.

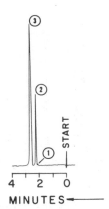

Figure 56. Analysis of inorganic gases [224]. Column:
75 ft × 1.0 mm ID open tubular coated with a thin molecular
sieve 5A layer. Carrier gas (He) flow rate at outlet: 12 ml/min.
Sample size: 10 μl gas. Column temperature: 26°C. Peaks:
1 hydrogen, *2* oxygen, *3* nitrogen.

at room temperature. Similarly, Purcell[40] obtained a very rapid
analysis of light hydrocarbon mixtures on copper open tubular
columns coated with a thin silica gel layer. Figure 55 shows one of
his chromatograms. Purcell was also successful in coating copper
columns of 1 mm ID with molecular sieve 5A flour; Figure 56 shows
the application of such a column for the separation of hydrogen,
nitrogen, and oxygen [224].

Kirkland [170] introduced a new material—fibrillar (crystalline)
colloidal boehmite with a very high (275 m^2/g) specific surface area—
for column coating. Relatively short (25 ft) columns of 0.25–0.50 mm
ID allowed the separation of Freon-type fluorinated hydrocarbons at
room temperature in less than two minutes.

In all the above-mentioned work the dynamic coating method
was utilized (see chapter 3.22). On the other hand, in the extensive
work of Halász and Horváth [123, 141] the static coating method
(see chapter 3.23) was applied.

In the studies of Halász and Horváth, graphitized carbon black
with a specific surface area of about 70 m^2/g was used most frequently
as the adsorbent. The powdered material, with an average particle
diameter of 1 μ or less, was dispersed in a high-density organic liquid
such as trifluorotrichloroethane or carbon tetrachloride (or their
mixtures). The column tubing was filled with this emulsion, and then
the solvent was evaporated as described in chapter 3.22.

---

[40] J. Purcell (Perkin–Elmer Corp., Norwalk, Conn.), unpublished results.

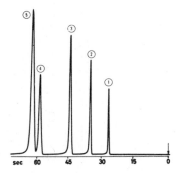

Figure 57. Analysis of benzene homologs [123, 141]. Column: 15 meter × 0.25 mm ID open tubular coated with a thin graphitized carbon back layer. Carrier gas (H_2) inlet pressure and flow rate at outlet: 1.7 atm (gauge), 12 ml/min. Sample size: 0.2 μl (liquid), split 1/1500. Column temperature: 245 °C. Peaks: 1 benzene, 2 toluene, 3 ethylbenzene, 4 m-xylene, 5 o- p-xylene.

The advantages of graphitized carbon black are [141] its high-temperature stability and good chemical resistance, its large homogeneous and apolar surface area, and the possibility of producing very fine particles. Open tubular columns coated with this material were successfully applied for the rapid separation of various paraffinic and aromatic hydrocarbons, and also of some polar substances such as alcohols and esters. Figures 57 and 58 show—according to Halász and Horváth [123, 141]—two chromatograms obtained on such columns. It is interesting to note that here *ortho-* and *para-*xylene will overlap, while *meta-*xylene is separated, which is contrary to the general experience with nonspecific partition columns.

Halász and Horváth also tried to coat the column tubing with various other adsorbents such as Al_2O_3, silicon carbide, and some organic pigments, but these columns showed relatively poorer performance. The most interesting results were obtained with the alumina-coated columns: it was found that by using hydrogen as

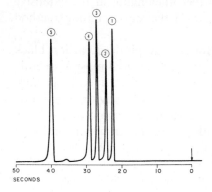

Figure 58. Analysis of some polar substances [123, 141]. Column: as in Figure 57. Carrier gas (H_2) inlet pressure and flow rate at outlet: 1 atm (gauge), 7.7 ml/min. Sample size: 0.2 μl (liquid), split 1/650. Column temperature: 200°C. Peaks: 1 methanol, 2 isopropanol, 3 diethyl ether, 4 ethyl acetate, 5 n-hexane.

carrier gas and working above 100°C, the active adsorbent layer actually behaved like a catalyst, and unsaturated hydrocarbons (e.g., ethylene, propylene, butene) were saturated by the hydrogen.

3.54 Modification of the van Deemter–Golay Equation for Open Tubular Adsorption Columns

Most of the expressions characterizing the gas chromatographic process can be applied for both partition and adsorption chromatography. However, the van Deemter–Golay equation as discussed in the second part of this book cannot be applied without some modifications, since there is now no liquid phase present and thus the C_L term is meaningless.

Recently, Giddings [105] investigated in detail the theory of gas adsorption chromatography and the modifications necessary in the basic equation expressing column performance. As derived by him, the C_L term has to be replaced by a mass transfer term for the kinetics of adsorption and desorption (C_K):

$$\text{HETP} = \frac{B}{\bar{u}} + (C_G + C_K)\bar{u} \tag{62}$$

If all sites on the adsorption surface are assumed to be equal, the C_K term can be expressed as follows:

$$C_K = \frac{8}{a\bar{u}_m}\left(\frac{k-1}{k}\right)^2 \frac{V_G}{S} \tag{63}$$

In Eq. (63), k is the partition ratio, V_G the total gas volume in a given column volume, and S the total surface area in a given column volume. The symbol $\bar{u}_m$ expresses the mean molecular velocity, and a is the so-called accommodation coefficient, which is the parameter commonly used to characterize the rate of adsorptive kinetics, equal to the fraction of molecules which stick to the surface upon impact. As given by Giddings, the values of a lie generally in the range from 0.01 to unity, while the mean velocity of the component molecules ($\bar{u}_m$) is approximately equal to the speed of sound, i.e., is of the order of 10^4 cm/sec.

Fourth Part

The Gas Chromatographic System

In the earlier chapters concerning the theory and practice of open tubular columns, the column was considered as a separate unit, independent of the other components of the gas chromatograph. But, of course, columns cannot be used alone, as an independent unit. One must rely on other components for proper sample introduction, adequate heating of the column, and sensitive detection and accurate recording of the separated peaks. In other words, the column has to be incorporated into a gas chromatographic system.

Generally, there is no basic difference in the functions and characteristics of the various parts of this system regardless of whether open tubular or packed columns are used. Much information concerning them has already been provided in the various excellent textbooks on gas chromatography, and there is no reason to repeat it here. However, in addition to these general characteristics, there are some additional specifications and requirements which apply specifically to the use of open tubular columns. In this connection, one should never forget that although the performance of the column can—and should—be treated separately, the *apparent* column performance—and this is ultimately important for the practical gas chromatographer—is dependent on whether the auxiliary devices in the system allow the full utilization of the original column performance.

In this part of the book, special requirements are summarized which must be considered when open tubular columns are employed in the usual gas chromatographic systems. Also, some special techniques are described which are related to the operation of the entire system.

4.1 SAMPLE INTRODUCTION

Chapter 2.5 discussed the fact that most of the open tubular columns used in practice have a very limited sample capacity. The sample size which can be introduced into these columns without overloading is usually of the order of 10^{-2}–10^{-3} μl of liquid. It is evident that the introduction of such small sample volumes by conventional techniques is practically impossible: the smallest microsyringes used generally for sample injection have a capacity of 0.5 μl, and volumes of less than 0.1 μl cannot be introduced reproducibly with these syringes.

In order to overcome this difficulty, an indirect sampling procedure may be used [3]; a relatively large (0.5–1 μl) liquid sample is injected into the system, evaporated, and either the sample vapor or its homogeneous mixture with the carrier gas is split into two very unequal parts, the smaller of which is introduced by the carrier gas flow into the column while the larger is discarded. In this way, the actual sample entering the column will be only a fraction of the originally injected volume.

4.11 Sample Introduction Systems

Since this book deals only with open tubular columns and not with gas chromatography in general, problems of the sample introduction systems not related specifically to this type of column are not discussed. However, three criteria for an adequate sample introduction system must be emphasized:

(a) A basic criterion of any gas chromatographic sampling system is that the liquid sample be introduced very sharply and evaporate instantaneously and completely. Furthermore, care has to be taken to avoid any possibility of back-diffusion of the evaporated sample.

(b) The temperature of the vaporizer part of the sample introduction system should be controlled and uniform. Too low a temperature results in slow evaporation and, thus, in fractionation of the sample; a too hot or not uniformly heated vaporizer will result in thermal breakdown of the injected sample or sample vapor. Also, the carrier gas should be preheated before coming into contact with the sample.

(c) A further prerequisite of a proper sample introduction device is that the sample and carrier gas be mixed homogeneously and

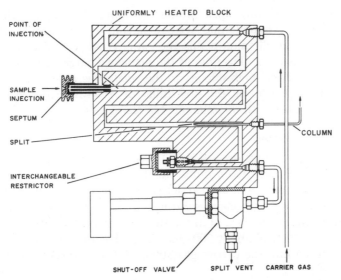

Figure 59. Schematic of a sample injection block. (Model 226 gas
chromatograph of The Perkin–Elmer Corp., Norwalk, Conn.)

that a *sharp* sample plug with uniform concentration along its
length travel to the split point and column. This criterion is
particularly critical because there must be sufficient volume to
achieve a homogeneous mixture, but the volume must not be
excessive, for this could lead to undesired spreading of the plug
before splitting or before entering the column.

When open tubular columns were first used, the present microsyringes were not yet available. Therefore, early workers—e.g., Desty
et al. [66], Halász and Schreyer [127, 239]—designed and built
ingenious devices for the reproducible and instantaneous injection of
small (0.5 μl) volumes of liquid samples. The introduction of precision
microsyringes,[41] however, greatly simplified this part of the problem,
and today they are generally used for the injection of liquid samples
into a heated block.

These injection blocks must be constructed carefully in order to
take all the above-mentioned criteria into consideration. As an
example, Figure 59 shows the simplified schematic of the sample
injection system of a commercial gas chromatograph designed

---

[41] For example, by the Hamilton Company, Whittier, California.

especially for use with open tubular columns. The liquid sample is injected with a hypodermic syringe having a long needle; during injection, the needle almost fills the entire cross section of a tube in the carrier gas path. The flow of the preheated carrier gas is directed so that it has a very high velocity in this small tube, around the syringe needle, thus preventing the possibility of back-diffusion. The vent flow is regulated with interchangeable restrictors.

4.12   Split Systems

The split systems described in the literature and used in practice can be divided into two groups. In the first group a homogeneous mixture of sample vapor *and* carrier gas moving continuously through the system is split (*dynamic splitting*), while in the second case the sample is vaporized in a static system and a part of its vapor (the volume of which is about 200–300 times that of the liquid sample) is introduced into the carrier gas stream flowing toward the column (*static splitting*).

In order to be effective, the stream splitter must be completely nondiscriminatory, i.e., it must divide all components of the sample mixture in the same ratio. If this is the case, the split system is called "linear." As pointed out elsewhere by the author [80, 84], a proper splitting device must fulfill three criteria:

(a) The relative sizes of peaks from a sample mixture must be identical to either calculated values or to the relative areas found with the same detector without splitting.
(b) In analyses of mixtures with different concentrations, the peak area values must be proportional to the concentrations.
(c) The relative sizes of peaks must remain constant even when analytical conditions such as temperature, split ratio, flow rates, etc., are varied.

This means that it is not enough to demonstrate only that the peak area values are reproducible or that peak area is a linear function of sample concentration under constant operating conditions, since the peak area values could also remain the same if the split were nonlinear with respect to fractionation because this fractionation could be *relatively* the same at different concentrations. Similarly, it is not enough to show only that a change in the operating parameters (e.g., in the split ratio or sample size) has no influence on the relative peak area values because it is possible that conditions outside the split system have brought about a change of the relative peak areas. For example, partial decomposition during evaporation might alter the effective composition of the sample.

4.121 Dynamic Splitting. In this case, a homogeneous mixture of the sample vapor and carrier moves through the split system, where it is split into two highly unequal parts, the smaller being introduced into the column itself while the larger is vented. If the sample and carrier gas are truly homogeneously mixed prior to splitting, the sample will be split in the ratio determined by the two flow rates.

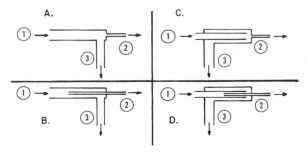

Figure 60. Various split systems. *1* Carrier gas + sample plug, *2* open tubular column, *3* vent (to pneumatic restriction).

The construction of the various dynamic split systems is shown in Figure 60. At the beginning, simple tee fittings (Figure 60a) were used; however, it was soon found that they often did not behave linearly. Today, practically all split systems used in practice consist of two concentric tubes (Figure 60b), and the linearity of such systems has been proved repeatedly [80, 84, 88]. Halász and Schneider [125, 126] described a modification of this general split system, claiming a better flow pattern for its branching section (Figure 60c). Another modification (Figure 60d) was reported by Clarke [53].

In all these systems, a pneumatic restrictor is placed in the vent line, and the split ratio (at a given inlet pressure) is dependent upon the relative pneumatic impedances of the concentric capillary tube plus the column on the one side and the restrictor in the vent line on the other. Therefore, in order to obtain a constant pneumatic resistance ratio, the resistance of the restrictor in the vent line must be extremely constant during operation. Both needle valves and (interchangeable) capillary tubes of various diameters have been used, the latter being somewhat superior with respect to constancy and reproducibility.

In the sample introduction system shown in Figure 59, the split is an integral part of the whole system. The pneumatic impedance is provided by interchangeable needles of various internal diameters. Downstream of this restrictor, a shutoff valve is incorporated. There are two reasons why it is advisable to have the vent line closed during analysis and open it only for a short time for the sample introduction period. First, in this way one can save a fairly large amount of carrier gas; more important, however, in this way a constant carrier gas flow through the column can be more easily maintained. Kaiser [160] pointed out that it is advantageous to maintain a very small flow of carrier gas through the "vent" line even when the shutoff valve is closed, in order to avoid any possibility of back-diffusion. This can be done either by a very-small-diameter capillary tube bypassing the valve or by actually constructing a "leak" in the valve. In the system of Figure 59, the latter principle is followed.

4.122 Static Splitting. In the second type of split system, the sample is first evaporated in a static system and then a small part of this homogeneous vapor is introduced into the carrier gas flowing toward the column. Such a system was described, for example, by Fejes *et al.* [94, 95]; a schematic is shown in Figure 61. In this system, a micropipet is filled with the liquid sample and then pushed halfway down in a channel. The vaporization chamber of known

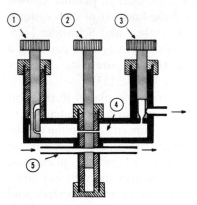

Figure 61. Split system for vaporized samples [94, 95]. *1* Pipet for the introduction of the liquid sample, *2* split device, *3* needle valve, *4* evacuated evaporation chamber, *5* carrier gas line.

volume is then evacuated through a needle valve, which is then closed. Then the micropipet is pushed down and the liquid evaporates and fills the chamber. By pushing down rod *2* only the vapor in the sample volume of this rod (which is a very small portion of the volume of the whole chamber) is introduced into the carrier gas line.

4.2 THE PNEUMATIC SYSTEM

The pneumatic system of a gas chromatograph can be divided into three parts: the carrier gas pressure or flow regulation, the column, and the connecting tubes between the various parts of the gas chromatograph through which carrier gas is flowing. The column itself and its characteristics regarding the carrier gas flow were discussed previously. However, some special problems associated with the two other parts of the pneumatic system should also be considered.

4.21 Carrier Gas Regulation

In isothermal gas chromatography, a constant carrier gas flow rate is maintained through the separation column. In the case of packed columns of the usual (2–4 mm) internal diameter, the constant flow rate is maintained by regulating either the inlet pressure or the flow rate itself. There are a large number of commercially available pressure and flow regulators which can be used for this purpose. These flow regulators perform well if the carrier gas flow rate is maintained at a reasonable level (above about 10 ml/min).

It has been stated that the carrier gas flow rate with small-diameter open tubular columns is relatively low, usually of the order of 0.5–5 ml/min. The regulation of such small volumetric flow rates is practically impossible with standard automatic flow regulators. Therefore, with such columns, one is generally restricted to the regulation of the carrier gas inlet pressure.

For work under isothermal conditions, this fact does not represent a particular problem since a constant inlet pressure (i.e., a constant pressure drop along the column) would automatically maintain a constant volumetric flow rate (assuming, of course, that the outlet pressure remains constant). However, when the temperature of the column is programmed while the pressure drop is kept constant, the volumetric carrier gas flow rate will not remain constant but will change considerably. Figure 62, for example, shows the plot of carrier gas flow rate *vs.* column temperature for a 150-ft-long open

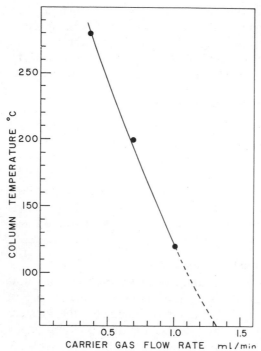

Figure 62. Plot of carrier gas flow rate at column outlet against column temperature for a constant 20 psi (gauge) pressure drop through an open tubular column. Column dimension: 150 ft × 0.25 mm ID.

tubular column with 0.25 mm nominal internal diameter, using helium as carrier gas and maintaining a constant pressure drop of 20 psi (gauge) through the column.

The basic reason why the volumetric flow rate is kept constant during analysis when programming the temperature of packed columns is that the thermal conductivity detectors normally used with commercial gas chromatographs are affected by changes in flow. It should be pointed out, however, that in such systems the average linear gas velocity does not remain constant even though the volumetric flow rate is kept constant. The reason for this is that a continuous change in the inlet pressure is necessary to keep a constant volumetric flow rate; but changing the inlet pressure also changes the pressure correction factor used for the calculation of the average linear gas velocity.

Golay *et al.* have suggested [114] that, from the theoretical point of view, *isobaric operation* (i.e., maintaining a constant pressure drop rather than a constant flow rate) has many advantages in programmed-temperature analysis. Naturally, in this case, detectors not affected by carrier gas flow changes have to be used in order to make quantitative analysis possible. As demonstrated by the author [88], flame ionization detectors are generally unaffected by the carrier gas flow changes involved.

The utilization of isobaric operation in programmed-temperature analysis with open tubular columns will be discussed in chapter 4.3.

A third possibility of carrier gas regulation has recently been mentioned in two short papers.[42,43] In this case, the *carrier gas flow rate or inlet pressure is programmed during analysis.* The basic advantage of such systems would be an increase in the admissible sample size. To date no experimental data on open tubular columns have been published.

4.22 Connecting Tubes

It has already been pointed out that spreading of the sample plug in the carrier gas stream must be avoided to maintain high column performance. In work with open tubular columns, the peaks are generally very sharp, i.e., their width is small. Any unnecessary large volume in the carrier gas line would result in spreading of the traveling sample plug, thus widening the peak widths and lowering the apparent column performance.

These "dead volumes" are particularly critical in the case of open tubular columns since here the sample volumes and the carrier gas flow rates are much lower than those used with packed columns. For example, let us suppose a connecting tube of 50 mm × 1 mm ID between column and detector; its volume is 39.3 μl. With a packed column and a carrier gas flow rate of 50 ml/min, the column effluent will pass through this tube in 0.047 second. However, with an open tubular column and a flow rate of 1 ml/min, the time of passage will be 50 times higher, i.e., 2.35 seconds. It is evident that the latter can represent a significant part of the peak width, while in the case of the packed column the time of passage is insignificant as compared to the peak width. Where dead volumes are of such geometry that laminar flow is interrupted and a "mixing volume" is created, this peak broadening effect will become even more significant, in addition to which peak "tailing" will also be evident.

---

[42] R. P. W. Scott, *Nature* **198**:782 (1963).
[43] S. A. Clarke, *Nature* **202**:1106 (1964).

In conclusion, one can say that in the design of gas chromatographs, special consideration is necessary if the use of open tubular columns is desired, and that the length and volume of the tubes connecting the column with the sampling system and the detector must be kept as small as possible. Usually, the dead volume *after* the column is more critical, and in some designs a secondary (scavenger) carrier gas flow is provided in order to reduce the passage time of the column effluent.

4.3   COLUMN HEATING
4.31   Isothermal Operation

Uniform heating of the separation column is a crucial problem in any gas chromatograph. There are two general requirements in this respect:

(a) The temperature should be controlled and measured fairly accurately.
(b) The temperature of the *entire* column should be uniform.

Today, most commercial gas chromatographs utilize an air bath oven in which hot air is circulated with the help of a high-speed blower. There is no specific difference in the requirements for packed and open tubular columns. One should, however, not forget that the latter may represent a higher mass if thick-walled tubing is used and that the physical shape of the column should permit free circulation of air, in order to avoid the possibility of temperature differences along the column.

It is also possible to use liquid baths for isothermal heating of open tubular columns. This technique is particularly useful for operation below ambient temperature.

4.32   Programmed-Temperature Operation

In programmed-temperature operation, the temperature of the column is raised during analysis. Usually a linear program rate is followed, but temperature programming is not restricted to this one type of temperature profile.

The initial and final temperatures and the program rate define the theoretical program (see Figure 63). However, in practice the *actual* column temperature will always differ from this theoretical program, mainly because of heat transfer delay. Since most of the open tubular column configurations in use are large coils with

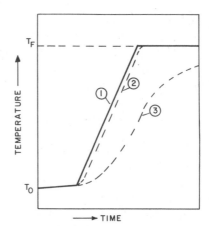

Figure 63. Correlation of temperature and time in programmed-temperature operation. *1* Ideal program, *2* desired actual program, *3* undesired program. T_0 and T_F = initial and final isothermal column temperatures.

multiple layers, care must be taken to provide a free air path to the inner tubing layers. In addition, it is very important that the air circulation in the oven be properly designed. McEwen [195] demonstrated clearly that with poor design—even with properly coiled columns—actual temperature curves may be obtained like that shown, e.g., in Figure 63, which are, of course, unacceptable.

The difference between theoretical and actual program is usually given as the time lag of the column temperature behind the theoretical program. Naturally, this lag depends on the program rates used and will increase when higher rates are used. In properly designed instruments, it can be less than 30 seconds for program rates up to about 30°C/min. If closer agreement is desired, the usual air bath oven is fundamentally inadequate.

Gill and Averill [106] in 1962 described a very special heating system in which the lag can be minimized.[44] In this system thin-walled tubing is used and wound into a flat spiral mounted between two thin circular aluminum discs. This assembly is pressed evenly onto a heater cast into an aluminum disc of similar size in order to ensure good thermal contact. The entire unit is surrounded by insulating bricks. In work with this unit and a program rate of 50°C/min, the measured lag at 300°C was only 9.2 seconds.

As shown in chapter 4.21, when open tubular columns are used (isobaric operation) one usually regulates the carrier gas inlet pressure

---

[44] This system is incorporated into the Model 226 gas chromatograph of the Perkin–Elmer Corporation, Norwalk, Conn.

and not the flow rate. As a result, in programmed-temperature operation, the flow rate of the carrier gas will change continuously.

According to Golay *et al.* [114], this mode of operation is advantageous with respect to two important characteristics of the mobile phase: the pressure drop along the column and the average linear gas velocity. If automatic flow regulators are used during temperature programming, both values will change continuously; however, in isobaric operation, only the latter changes. This means that isobaric operation considerably simplifies the mathematical treatment of the separation procedure. As a result, prediction of the elution time or the necessary program becomes possible.

In addition to the use of isobaric operation, temperature programming of open tubular columns differs from that of packed columns in the general rule that in work with open tubular columns lower program rates and lower initial temperatures are usually preferred. The theoretical considerations which lead to these rules have been discussed by Habgood and Harris [120]. It is generally not immediately recognized that even very low program rates can significantly speed up the analysis time, without loss of resolution. Habgood and Harris demonstrated this fact by using the most remarkable analysis of the gasoline fraction of Ponca City crude oil reported by Desty *et al.* [65]. This analysis was carried out at 25°C with a 270 meter × 0.152 mm ID open tubular column coated with squalane, and in 1240 min, 122 components up to *n*-nonane were separated. According to their calculation, with a program rate as low as 0.1°C/min starting at 25°C, the retention time of *n*-nonane could be reduced to 390 minutes while still maintaining overall column performance.

4.33 Applications of Temperature-Programmed Open Tubular Columns

Temperature-programmed open tubular columns are particularly suited for the analysis of samples with a wide range of boiling points. The advantages of such operation are obvious and have been discussed in detail in various general textbooks on gas chromatography.

The following few chromatograms are intended to illustrate the performance of temperature-programmed open tubular columns in the analysis of some complex samples. Most of these samples are actually natural mixtures, and they cover a wide variety of

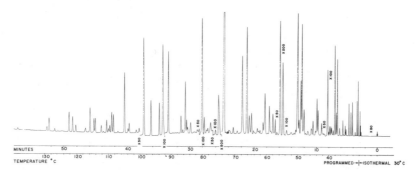

Figure 64. Analysis of an "86 Octane Powerformate" gasoline [106]. Column: 150 ft × 0.25 mm ID open tubular coated with DC-550 phenyl silicone oil liquid phase. Column temperature as given; program rate: 2°C/min.

applications. In all cases, isobaric operation was followed, i.e., the inlet pressure and not the flow rate of the carrier gas was held constant during analysis.

As in chapter 2.36, where the application of isothermally heated open tubular columns was shown, we begin here with natural hydrocarbon mixtures, which represent the most complex natural samples. Figure 64 illustrates the analysis of an "86 Octane Powerformate" gasoline [106]. Columns with silicone oil liquid phase and low program rates are highly suitable for the analysis of such complex hydrocarbon mixtures. Another petroleum product was analyzed for Figure 65, which shows the chromatogram of a diesel oil [166].

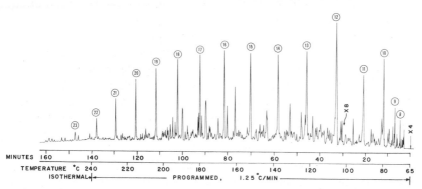

Figure 65. Analysis of a diesel oil [166]. Column: 50 meter × 0.25 mm ID open tubular coated with OS-138 poly(phenyl ether) liquid phase. Column temperature as given; program rate: 1.25°C/min. Peaks 8–23 correspond to the C_8–C_{23} normal paraffins.

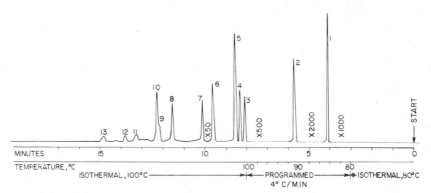

Figure 66. Analysis of benzene homologs. Column: 150 ft × 0.25 mm ID open tubular coated with *m*-bis(*m*-phenoxyphenoxy)benzene + Apiezon L liquid phase (8/2, w/w). Column temperature as given; program rate: 4°C/min. Peaks: *1* benzene, *2* toluene, *3* ethylbenzene, *4* *p*-xylene, *5* *m*-xylene, *6* *o*-xylene, *7* isopropylbenzene, *8* *n*-propylben- zene, *9* *p*-ethyltoluene, *10* *m*-ethyltoluene, *11* 1,3,5-trimethylbenzene, *12* *o*-ethyltoluene, *13* 1,2,4-trimethylbenzene.

Here, the peaks of normal paraffins which constitute the largest part of the sample are identified. Again a very slow program rate was applied.

Coal tar fractions are rich in aromatic hydrocarbons. The analysis of the benzene homologs under isothermal conditions has already been shown in Figure 5, which demonstrated the high per- formance of the *m*-bis(*m*-phenoxyphenoxy)benzene liquid phase in the separation of closely related isomers. When used for the analysis of benzene homologs, this substrate is usually mixed with a nonpolar liquid phase because it alone would not separate ethylbenzene from *p*-xylene. In the case of Figure 5, squalane was used as the second com- ponent of the liquid phase. However, if programmed-temperature operation of the column is desired, the addition of Apiezon oil instead of squalane is recommended because squalane has a high bleed rate (see chapter 4.34). Figure 66 demonstrates the analysis of a synthetic mixture similar to that analyzed in Figure 5, but now using programmed-temperature operation.[45] As seen, except for the *m*/*p*-ethyltoluene pair, all other components could be separated. It should further be mentioned that if higher condensed ring aromatics are to be analyzed, *m*-bis(*m*-phenoxyphenoxy)benzene alone, without

---

[45] E. W. Cieplinski (Perkin–Elmer Corp., Norwalk, Conn.), unpublished result.

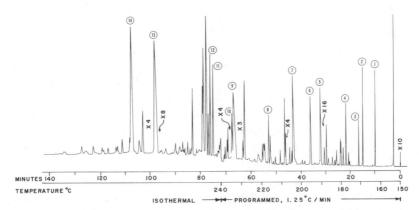

Figure 67. Analysis of anthracene oil [166]. Column: 50 meter × 0.25 mm ID open tubular coated with OS-138 poly(phenyl ether) liquid phase. Column temperature as given; program rate: 1.25°C/min. Identified peaks: *1* naphthalene, *2* β-methyl-naphthalene, *3* α-methylnaphthalene, *4* biphenyl, *5* acenaphthene, *6* biphenylene oxide, *7* fluorene, *8* dihydrophenanthrene, *9* phenanthrene, *10* anthracene, *11* acridine, *12* carbazole, *13* fluoranthene, *14* pyrene.

the addition of a nonpolar phase, can also give excellent separation in many cases.

The analysis of an anthracene oil fraction is shown in Figure 67. This natural sample consists of a large variety of condensed ring aromatics, some of them identified. Again a low program rate of

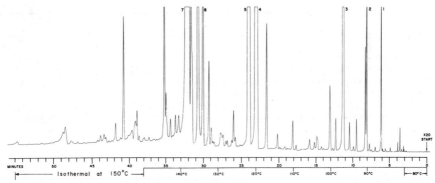

Figure 68. Analysis of a natural peppermint oil sample from Yakima Valley, Washington [13, 50]. Column: 150 ft × 0.25 mm ID open tubular coated with Ucon oil 50 HB 2000 liquid phase. Carrier gas (He) inlet pressure: 20 psi (gauge). Column temperature as given; program rate: 2°C/min. Identified peaks: *1* α-pinene, *2* β-pinene, *3* eucalyptol, *4* menthone, *5* menthofuran, *6* menthyl acetate, *7* menthol.

1.25°C/min was applied. Note particularly the successful separation of phenanthrene and anthracene.

Following petroleum and coal tar fractions, essential oils represent the second natural sample groups which show the highest complexity. Figure 68, for example, shows the analysis of a peppermint oil sample, based on the work of Cieplinski and Averill [13, 50]. Here, low attenuation—i.e., high sensitivity—was used in order to demonstrate the multitude of minor constituents. Some of the main components were identified; however, further identification would represent a very complicated problem.

Teranishi and co-workers of the Western Research Laboratory of the U.S. Department of Agriculture pioneered in the identification of the individual components in essential oils utilizing open tubular columns in connection with mass spectrometry. Figure 69 is one of their chromatograms showing the analysis of a "Gebrig" hop oil [40, 265]. In this case, a 100 ft × 0.25 mm ID open tubular column coated with Tween 20 liquid phase was used in connection with a flame ionization detector. For structure identification, part of the

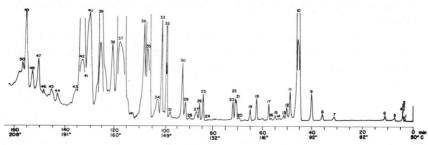

Figure 69. Analysis of a "Gebrig" hop oil [40, 265]. Column: 100 ft × 0.25 mm ID open tubular coated with Tween 20 (polyoxyethylene sorbitan monolaurate) liquid phase. Carrier gas (He) inlet pressure: 12 psi (gauge). Sample volume: 4 μl (liquid), split 1/300. Column temperature as given; nonlinear program rates. Peaks: *1* n-pentane*, *2* pentene-2 or isomer, *3* diethyl ether*, *4* isoprene*, *5* acetone*, *6* octane*, *7* α-pinene*, *8* 2-methylpropyl isobutyrate*, *9* myrcene*, *10* 2-methylbutyl isobutyrate*, *12* limonene*, *13* $C_{10}H_{16}$, *14* a methyl ester, *15* $C_{10}H_{16}$, *16* p-cymene*, *17* methyl heptanoate*, *18* methyl hept-4-enoate, *19* methyl-6-methylheptenoate, *20* 1-phenyl ethanol, *21* nonanone-2*, *22* methyl octanoate*, *23* $C_{10}H_{14}O$ or $C_{11}H_{18}$, *24* decanone-2, *25* methyl nonoate*, *26* linalöol*, *27* methyl octenoate, *28–31* $C_{15}H_{24}$ isomers, *32* undecanone-2*, *33* methyl dec-4-enoate, *34* $C_{15}H_{24}$, *35* β-caryophyllene*, *36* methyl dec-4,8-dienoate, *37* humulene*, *38–40* $C_{15}H_{24}$ isomers, *41* tridecanone*, *42* two $C_{15}H_{24}$ isomers, *43* $C_{15}H_{22}$, *44* $C_{15}H_{20}$, *45* tetradecanone-2, *46* $C_{15}H_{20}$, *47–48* $C_{15}H_{24}O$ isomers, *49* pentadecanone-2 + $C_{15}H_{24}O$ + $C_{15}H_{28}O$, *50* $C_{15}H_{24}O$ + $C_{15}H_{26}O$. Peaks marked with an asterisk (*) were identified both by comparison with pure standards and by mass spectrometer.

column effluent was led into the ionization chamber of a Bendix "Time of Flight" Mass Spectrometer, utilizing an auxiliary vacuum pump in order to maintain the desired pressure in the latter. Detection of a component in the effluent was made by observing the mass spectrum on an oscilloscope. The presence of certain components could be confirmed by injection of the authentic pure substance mixed with a sample of the hop oil; these components are marked in the caption of Figure 69 by asterisks. In the case of the other peaks, pure substances were not available, and thus the identity was predicted from the mass-spectral pattern by comparison with the characteristic pattern of homologs.

In the gas chromatographic system used for the analysis shown in Figure 69, the temperature of the column was programmed manually. This is the reason for the nonlinear program rates. Where the peaks went off-scale, successive attenuation was used to bring the peak on scale.

Of course, temperature-programmed open tubular columns need not always be used with low program rates. The combination of short columns and high program rates can often result in a very short analysis time. Figure 70, for example, shows—according to Marco [190]—the analysis of C_1–C_{12} alcohols on a 30 ft × 0.50 mm

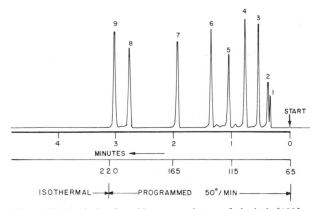

Figure 70. Analysis of a wide-range mixture of alcohols [190]. Column: 30 ft × 0.50 mm ID open tubular coated with Carbowax 20 M liquid phase. Carrier gas (N_2) inlet pressure: 4 psi (gauge). Column temperature as given; program rate: 50°C/min. Peaks: *1* methyl, *2* ethyl, *3* isopropyl, *4* isobutyl, *5* isoamyl, *6* isohexyl, *7* isooctyl, *8* isohendecyl, and *9* isododecyl alcohol.

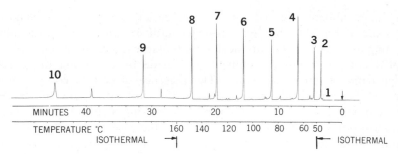

Figure 71. Analysis of the mixture of C_1-C_{12} normal alcohols [106]. Column: 150 ft × 0.25 mm ID open tubular coated with nitril and phenyl silicone oils in 50/50 ratio. Column temperature as given; program rate: 5°C/min. Peaks: *1* methyl, *2* ethyl, *3* propyl, *4* butyl, *5* amyl, *6* hexyl, *7* heptyl, *8* octyl, *9* decyl, and *10* dodecyl alcohol.

ID column with a program rate of 50°C/min. It will be seen that isododecyl alcohol emerged in 3 minutes. As a comparison, Figure 71 (from the work of Averill [106]) illustrates the analysis of a similar mixture but with a longer and narrower column and with a low (5°C/min) program rate. Here the total analysis time was about 45 min; the separation of the individual components is naturally better. These two chromatograms illustrate well the flexibility of the

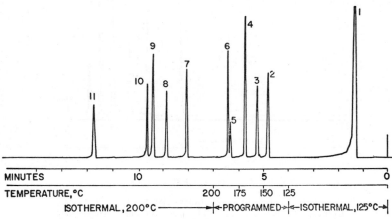

Figure 72. Analysis of N-trifluoroacetylamino acid butyl esters [106]. Column: 70 ft × 0.25 mm ID open tubular coated with FS-1265 fluorinated silicone oil liquid phase. Column temperature as given; program rate: 25°C/min. Peaks: *1* solvents, *2* alanine, *3* glycine, *4* valine, *5* isoleucine, *6* leucine, *7* proline, *8* methionine, *9* aspartic acid, *10* phenylalanine, *11* glutamic acid derivates.

application of temperature-programmed open tubular columns and the influence of column length, diameter, and program rate on the separation and analysis time.

Another application of a high program rate is shown in Figure 72. Here, a mixture of some N-trifluoroacetylamino acid butyl esters was analyzed [106]. The reason for the long initial isothermal temperature was to maintain the column temperature relatively low during the transit of the first peaks; on the other hand, a quick programming was necessary in order to be able to elute the later peaks in a short time. A slower program rate would undoubtedly result in a better separation between peaks but, at the same time, would result in a very long analysis time.

Finally, Figure 73 shows—according to Grob [119]—the analysis of some strong bases on an open tubular column. The separation of such substances by gas–liquid chromatography has long been a serious problem, and although packed columns with specially treated support material can be used, their separation ability is limited. On the other hand, open tubular columns with a special liquid phase consisting of a mixture of polyethylene imine (MW = 800) and silanized Carbowax 1000 showed an excellent separation of the

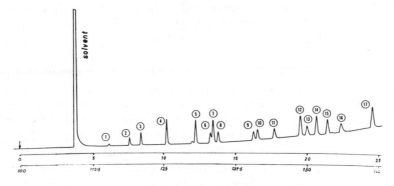

Figure 73. Analysis of strong organic bases [119]. Column: 50 meter × 0.50 mm ID open tubular coated with a mixture of poly(ethylene imine) and silanized Carbowax 1000 (2/5, w/w). Column temperature as given; program rate: 2.5°C/min. Carrier gas (N₂) inlet pressure: 0.3 atm (gauge). Peaks: *1* pyrrolidine, *2* piperidine, *3* pyridine, *4* 2-methylpyridine, *5* 2,6-dimethylpyridine, *6* 2-ethylpyridine, *7* 3-methylpyridine, *8* 4-methylpyridine, *9* 2,5-dimethylpyridine, *10* 2,4-dimethylpyridine, *11* 2,3-dimethylpyridine, *12* 2,4,6-trimethylpyridine, *13* 2,3-trimethylpyridine, *14* 4-ethylpyridine, *15* 3,5-dimethylpyridine, *16* 3,4-dimethylpyridine, *17* pyrrole.

many pyridine homologs. Figure 73 shows the chromatogram of a complex mixture of organic bases.

4.34 Compensation of Liquid-Phase Bleeding During Temperature Programming

Every liquid phase used in gas chromatography has a certain vapor pressure at a given temperature, and because the gas in the column is constantly moving, the column effluent always contains a certain amount of its vapor. This phenomenon is called bleeding of the liquid phase. Since gas chromatographic detectors usually respond to any substance other than the inert carrier gas, substrate bleeding will result in a detector background signal, which has to be suppressed.

If column temperature and flow rate are unchanged during analysis—in isothermal operation—the rate of bleeding is constant. Thus, the background signal due to substrate bleeding is also kept constant and can be compensated electrically.

The situation is different in programmed-temperature operation. Here, the column temperature is continuously raised. Since the vapor pressure of any substance increases exponentially with temperature, the rate of liquid phase bleeding will increase similarly during the program. As a result, the background signal will not remain constant any more but will rise correspondingly. This in turn, will cause an upward drift of the baseline during analysis.

In order to overcome this difficulty, a special technique was introduced in 1961 by Emery and Koerner[46] for the elimination of baseline drift when the temperature of the column is programmed. In this technique, a reference column identical to the analytical column and exposed to the same temperature environment is used to compensate for liquid phase elution. In the original system of Emery and Koerner, the effluent of the reference columns was directed to the reference cell of a hot-wire thermal conductivity detector, while the analytical column effluent was directed to the sensing cell of the detector. Since the output of the detector bridge is a measure only of the differences in thermal conductivity of the two chambers and the liquid phase elutes equally from both columns, they compensate each other and the output of the bridge will be zero except when sample components enter the sensing cell.

---

[46] E. M. Emery and W. E. Koerner, *Anal. Chem.* **33**:523 (1961).

The dual-column mode was later extended to flame ionization detectors,[47] and in 1962 Condon developed a differential flame ionization detector which simplifies the operation of the gas chromatographic system [87]. Today, the dual-column technique—with both thermal conductivity and flame ionization detectors—has found wide acceptance in practice.

The bleeding of the liquid phase is of course independent of the type of column used. In practical analyses, however, it is generally less significant with open tubular than with packed columns. This can be clearly seen in the preceding chromatograms, where generally only a slight baseline drift can be observed. The main reason for this is that open tubular columns are generally used at lower temperatures than the corresponding packed columns and, as mentioned above, the bleeding rate increases exponentially with temperature. However, if necessary, open tubular columns can be adapted for dual-column use. The compensation of baseline drift is usually carried out not by flow adjustment but by using independent pressure regulation on each column and adjusting the inlet pressure of the reference column in order to obtain the same bleeding as observed from the column used for analysis. Of course, with open tubular columns of larger diameters, standard flow regulators can again be utilized.

Figures 74 and 75 illustrate the difference between uncompensated single-column and compensated dual-column operation for the analysis of commercial gasoline [87]. The columns used were 150 ft × 0.25 mm ID and coated with squalane. In both cases, 1 μl liquid sample was injected and split in a ratio of about 1/500.

4.4 DETECTION AND RECORDING

The use of open tubular columns creates a threefold problem regarding detection and recording:

(a) The flow rates used are generally fairly low; in most cases, they are less than 10 ml/min and flow rates of the order of 0.5–2 ml/min are common.

(b) In work with small-diameter ("capillary") open tubular columns, only very small samples can be analyzed. Actual sample sizes reaching the column are often in the range of $10^{-5} - 10^{-6}$ gram; thus, e.g., a component present in 1 % concentration in such a sample

[47] R. Teranishi, R. G. Buttery, and R. E. Lundin, *Anal. Chem.* **34**:1033 (1962).

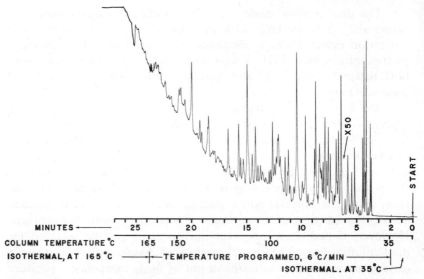

Figure 74. Analysis of a commercial gasoline sample. I. Single-column operation [87]. Column: 150 ft × 0.25 mm ID open tubular coated with squalane liquid phase. Column temperature as given; program rate: 6°C/min. With attenuation × 50, full-scale response corresponds to 1.25×10^{-10} ampere.

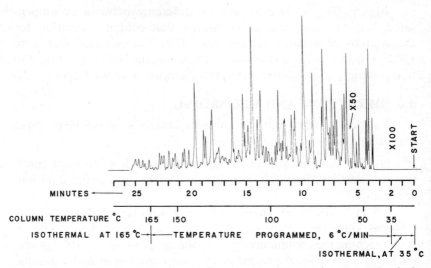

Figure 75. Analysis of a commercial gasoline sample. II. Dual-column operation [87]. Each column, conditions, and sample as given in Figure 74.

corresponds to $10^{-7} - 10^{-8}$ gram. Such very small amounts require detectors of high sensitivity.

(c) The early peaks—and in the case of high-speed analyses, all peaks—are very sharp and peak widths of less than a second are certainly possible. Thus, the response time of the detection and recording system has a much bigger influence on the true presentation of the results than in the case of packed columns. As pointed out by Schmauch,[48] the time constant of the whole detection and recording system must be less than one-twentieth of the peak elution time (base width) if the peak is to be truly recorded.

4.41 Commonly Used Detectors

For the reasons listed above, ionization detectors are generally used with open tubular columns. It was fortunate that the development of these highly sensitive detection devices coincided with that of the open tubular column. Among the ionization detectors, the flame ionization detector is definitely preferred by the practical user. Its basic advantage is that it has essentially zero volume; thus, in proper instrument construction, the column effluent can be directly mixed with hydrogen at the detector and no scavenger gas is required downstream of the column. Beta-ray (argon) ionization detectors have also been used frequently with open tubular columns; here, scavenger gas is added to the column effluent. The situation with the electron capture detector[49] is similar, although very little published data are yet available on its use in connection with open tubular columns [187a].

At the moment, the use of standard thermal conductivity detectors is restricted in practice to open tubular columns of larger diameter. The intrinsic sensitivity of these devices would be satisfactory in most cases, and although their time constant is slower than that of the ionization detectors, this again is not the basic problem, except, of course, for high-speed analyses. Golay, in his work, actually utilized specially constructed micro thermal conductivity detectors and obtained satisfactory chromatograms. Condon showed some chromatograms for which similar detectors were used at the 1960

---

[48] L. J. Schmauch, *Anal. Chem.* **31**:225 (1959).

[49] According to a private communication of Drs. A. Liberti (University of Naples) and G. P. Cartoni (University of Rome), they have successfully used open tubular columns with an electron capture detector adding about 99 ml/min scavenger gas to a 1 ml/min carrier gas flow downstream of the column.

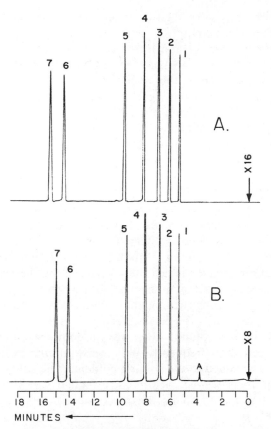

Figure 76. Chromatograms of a hydrocarbon mixture obtained using (a) a conventional flame ionization detector (FID) and (b) a specially constructed micro-thermistor detector (MTD) [230]. Column: 50 meter × 0.25 mm ID open tubular coated with squalane liquid phase. Sample size: 1 μl (liquid), split 1/400. Carrier gas (He) flow rate at outlet: 1.05 ml/min. Temperature of column and MTD: 60°C. Peaks: A air, 1 n-pentane, 2 2,2-dimethylbutane, 3 2,3-dimethylbutane, 4 n-hexane, 5 2,4-dimethylpentane, 6 isooctane, 7 n-heptane.

Eastern Analytical Symposium [55], and the detector used by Purcell [224] for the analysis of inorganic gases was also a small-volume (15–20 mm^3) thermal conductivity detector (see Figures 20 and 56). However, the principal problem is that most of the commercial thermal conductivity detectors have an excessive volume, making them unsuitable for such applications.

In recent years, various commercial micro thermistor and hot-wire detectors have been marketed by the Gow-Mac Instrument Company, Madison, New Jersey, with cavity volumes of 25–115 mm^3, and Camin *et al.* [42] proved the applicability of the micro hot-wire detectors with 0.50 mm ID open tubular columns. According to Petrocelli [218], the small-volume thermistor cell can be adapted with slight modifications of the original construction even for 0.25 mm ID columns.

Roedel [230] recently described specially constructed micro thermistor detectors having a gas volume of only a few cubic millimeters and compared their performance to that of a commercial flame ionization detector. Figures 76*a* and *b* compare two chromatograms obtained by using a 50 meter × 0.25 mm ID open tubular column coated with squalane liquid phase. A 1 μl liquid sample was injected in both cases and split in a ratio of 1/400. In the case of Figure 76*a*, a conventional flame ionization detector was used, while the chromatogram shown in Figure 76*b* was obtained using a specially constructed micro thermistor detector with a gas volume of only 2.5 mm^3. Both chromatograms were recorded on a standard 2.5 millivolt potentiometer recorder, the attenuation levels being registered on the chromatograms; as can be seen, there is only a difference of a factor of two.

4.42 Auxiliary Detectors

Gas chromatography is a separation technique and practically all commonly used standard detectors respond to all or to a large number of compounds. Thus, the establishment of what particular substance corresponds to a chromatographic peak is often a very difficult problem. This problem is discussed in detail in the available textbooks. One possibility is the use of auxiliary detectors which help in interpreting the chemical composition of the individual fractions separated by the column. In gas chromatography, spectro-photometers and mass spectrometers are often used for this purpose.

The use of IR and UV spectrophotometers is generally restricted by the very small sample sizes available. However, mass spectrometry seems to be an ideal method for this purpose, and a large number of papers describe the use of mass spectrometers—either as the principal or as auxiliary detectors—in connection with open tubular columns (see the subject index of the bibliography). The identification of the individual peaks of the chromatogram shown in Figure 69 was carried out by the combined use of open tubular columns and mass spectrometry.

4.43   Recording Systems

It has already been mentioned that sometimes the peaks obtained on open tubular columns are very sharp, having widths of only a few seconds or even less. In such cases, only very fast recording systems such as oscilloscope readout or galvanometer-type recorders with a full-scale pen speed of 0.1 second are recommended. A chromatogram obtained using oscilloscope readout was shown in Figure 19; the high-speed chromatograms of Figures 20 and 21 were recorded on fast pen galvanometer-type recorders.

The most commonly used potentiometer-type recorders are available in two types: with a full-scale pen speed of 0.25 and 1 second. Although the first has definite advantages, particularly for the recording of the early fast peaks, 1-second recorders have the most widespread use, and an overwhelming majority of the chromatograms shown in this book were obtained by using such recorders.

If the recorder is not fast enough, a peak with a smaller area than it should have will result. This problem is particularly serious in the analysis of wide-boiling-range mixtures containing some light components; if the composition of the sample is calculated from the chromatogram (either by calculating the area of the individual peaks from the chart or using electromechanical integrators connected to the recorder), a significant error may result. In this case, the proper method for quantitative analysis is to use an electronic integrator connected directly to the amplifier and not to the recorder: thus, the proper peak area values can be obtained and the recorder serves only as a qualitative indicator.

Bibliography

In the fall of 1961, on the occasion of the fifth anniversary of the Golay columns, the Perkin–Elmer Corporation published a special issue of the *Instrument News*. In this issue, a bibliography listing all publications and papers which dealt with the theory and applications of open tubular columns was published. This bibliography was complete up to the middle of 1961 and consisted of 81 references. It was felt that a similar but up-to-date bibliography would be useful in this book.

The present bibliography is practically complete up to July 1, 1964; it contains 288 references. All of the papers listed here deal explicitly with the theory and application of open tubular columns or report on work in which such columns were used for analysis. Theoretical papers which deal with gas chromatography in general are listed only if they also refer explicitly to open tubular columns. General textbooks are not included, except for one volume of Kaiser's book [160], which deals exclusively with open tubular columns.

It is interesting to note the yearly distribution of the publications:

1958	5	references
1959	14	,,
1960	25	,,
1961	35	,,
1962	57	,,
1963	112	,,
(first half)1964	40	,,

288 references

Papers presented at various meetings are listed separately only if they have not yet been published. For papers that were subsequently published, only the reference to the final printed publication and not to the original presentation is given. Preprints are listed only if the text was not subsequently published. Thus, for example, the preprints of the 1957 and 1959 International Gas Chromatography Symposia of the Instrument Society of America (the so-called Lansing Meetings) are not listed, but only the final proceedings published in 1958 and 1961, respectively.

Brochures dealing with applications of open tubular columns are listed if they are available to the general public; data sheets, however, are incorporated into this bibliography only if in addition to showing a chromatogram they also discuss some aspects of the particular application.

The general system of editing of this bibliography is the same as that used by the author in the above-mentioned preliminary bibliography on open tubular columns and in the *Bibliography on Gas Chromatography. II*, published in the Proceedings of the Second International Gas Chromatography Symposium of the Instrument Society of America.[50]

In the compilation of this bibliography, two sources were of great help:

(a) The *Gas Chromatography Abstracts*, edited by C. E. H. Knapman and sponsored by the Gas Chromatography Discussion Group. These "blue books" are published yearly. The 1958–1962 issues were published by Butterworths, London and Washington, D.C., and beginning with 1963 they are being published by the Institute of Petroleum, London (England).

(b) The cards of the *Gas Chromatography Abstracts Service*, issued weekly by the Preston Technical Abstracts Co., Evanston, Illinois.

The bibliography is followed by a brief subject index. This index was prepared based on the titles of the papers and is intended only to give a first orientation rather than the detailed subject of each publication.

---

[50] *Gas Chromatography*, ed. H. J. Noebels, R. F. Wall, and N. Brenner, Academic Press, New York, 1961, pp. 375–455.

1. Adlard, E. R., M. A. Khan, and B. T. Whitham, Application of Capillary Columns in the Study of the Thermodynamic Behaviour of Ethanol and Carbon Tetrachloride in Dinonyl Phthalate, in: *Gas Chromatography 1962*, ed. M. Van Swaay, Butterworths, Washington, D.C., 1962, pp. 84–101.

2. Adlard, E. R., and B. T. Whitham, Analysis of Petroleum Fractions by Subtractive Gas Chromatography, in: *Gas Chromatography*, ed. N. Brenner, J. E. Callen, and M. D. Weiss, Academic Press, New York, 1962, pp. 371–390.

3. Amy, J. W., and W. E. Baitinger, Gas Chromatography Sample Injection, in: *Lectures on Gas Chromatography 1962*, ed. H. A. Szymanski, Plenum Press, New York, 1963, pp. 19–31.

4. Andersson, C. O., R. Rhyhage, S. Ställberg-Steinhagen, and E. Stenhagen, Mass Spectrometric Studies. IX. Methyl and Ethyl Esters of Some Aliphatic α-Amino Acids, *Arkiv Kemi* **19**:405–416 (1962).

5. Anonymous, Fractograms on Golay Columns (in German), *Aus der Praxis— für die Praxis*, No. 12–GC, Bodenseewerk Perkin–Elmer & Co., Überlingen, 1960.

6. Anonymous, Five Years of Golay Columns, *Instrument News* **13**(1a):1–2 (1961).

7. Averill, W., New Chromatographic Techniques Simplify Flavor and Odor Analysis, *Instrument News* **12**(1):6–7 (1960).

8. Averill, W., Coating and Testing of Capillary Columns (Discussion), in: *Progress in Industrial Gas Chromatography, Vol. I*, ed. H. Szymanski, Plenum Press, New York, 1961, pp. 225–230.

9. Averill, W., Use of Gas Chromatography for Analysis of Odors, Flavors, and Air Pollution, in: *Progress in Industrial Gas Chromatography, Vol. I*, ed. H. A. Szymanski, Plenum Press, New York, 1961, pp. 31–36.

10. Averill, W., Columns with Minimum Liquid Phase Concentration for Use in Gas Liquid Chromatography, in: *Gas Chromatography*, ed. N. Brenner, J. E. Callen, and M. D. Weiss, Academic Press, New York, 1962, pp. 1–6.

11. Averill, W., Gas Chromatography of Free Fatty Acids Using Golay Columns, *J. Gas Chromatog.* **1**(1):22 (1963).

12. Averill, W., Analysis of Tar Acids with Golay Columns, *Gas Chromatography Applications*, No. GC–DS–001, The Perkin–Elmer Corporation, Norwalk, Conn., 1963.

13. Averill, W., Analysis of Peppermint Oils, *Gas Chromatography Applications*, No. GC–DS–004, The Perkin–Elmer Corporation, Norwalk, Conn., 1963.

14. Averill, W., Analysis of Chlorinated Biphenyls and Polyphenyls with Golay Column, *Gas Chromatography Applications*, No. GC–DS–011, The Perkin–Elmer Corporation, Norwalk, Conn., 1964.

15. Averill, W., and L. S. Ettre, Gas Chromatographic Analysis of C_1–C_4 Hydrocarbons with Open Tubular Columns, *Nature* **196**:1198–1199 (1962).

16. Bank, H. M., J. C. Saam, and J. L. Speier, Addition of Silicone Hydrides to Olefinic Double Bonds. IX. Addition of *sym*-Tetramethyldisiloxane to Hexene-1, -2, and -3, *J. Org. Chem.* **29**:792–794 (1964).

17. Beaven, G. H., P. B. D. De La Mare, E. A. Johnson, and N. V. Klassen, Kinetics and Mechanism of Aromatic Halogen Substitution. XII. Products of Chlorination of Fluorene in Acetic Acid, *J. Chem. Soc. (London)* **1962**(3):988–993.

18. Beerthuis, R. K., Some Applications of Gas Chromatography in Fat Chemistry (in Dutch), *Chem. Weekblad* **59**(35):469–479 (1963).

19. Behrendt, S., Infrared Microspectrometer Using Molecular Beam to Monitor Capillary Gas Chromatography Effluents, *Nature* **201**:70 (1964).
20. Bernhard, R. A., Separation of Terpene Hydrocarbons by Gas Liquid Chromatography Utilizing Capillary Columns and Flame Ionization Detection, *Anal. Chem.* **34**:1576–1579 (1962).
21. Bernhard, R. A., and R. Wrolstad, Essential Oil of *Schinus Molle*: The Terpene Hydrocarbon Fraction, *J. Food Sci.* **28**:59–63 (1963).
22. Bey, K., Analysis of Skin Fats from Worn Clothes (in German), *Fette, Seifen, Anstrichmittel* **65**:611–618 (1963).
23. Blanchard, K. R., and P. R. Schleyer, Quantitative Study on the Interconversion of Hydrindane Isomers by Aluminum Bromide, *J. Org. Chem.* **28**:247–248 (1963).
24. Brandt, W. W., The Column in Gas Chromatography, *Anal. Chem.* **33**(8):23A–31A (1961).
25. Brenner, N., and L. S. Ettre, Characteristics of the Capillary Gas Chromatograph and Its Application to Quantitative Analysis, *Acta Chim. Acad. Sci. Hung.* **27**:205–214 (1961).
26. Brown, H., and G. Zweifel, Hydroboration. IX. Hydroboration of Cyclic and Bicyclic Olefins—Stereochemistry of the Hydroboration Reaction, *J. Am. Chem. Soc.* **83**:2544–2551 (1961).
27. Brown, R. A., and E. R. Quiram, Analysis of *o*-Xylene Oxidation Products by Combined Gas Chromatographic and Spectroscopic Techniques, *Appl. Spectry.* **17**(2):33–36 (1963).
28. Bruderreck, H., W. Schneider, and I. Halász, Quantitative Gas Chromatographic Analysis of Hydrocarbons with Capillary Columns and Flame Ionization Detector, *Anal. Chem.* **36**:461–473 (1964).
29. Bruner, F., and G. P. Cartoni, Gas Chromatographic Separation of Benzene and Deuterobenzenes, *J. Chromatog.* **10**:396 (1963).
30. Bruner, F., and G. P. Cartoni, Use of Glass Capillary Columns with Modified Internal Area in Gas Chromatography, *Anal. Chem.* **36**:1522–1526 (1964).
31. Bruner, F., G. Cartoni, and A. Liberti, Evaluation of Glass Capillary Columns in Gas Phase Chromatography (in Italian), *Chim. Ind. (Milan)* **44**:999–1001 (1962).
32. Brunnée, C., L. Jenkel, and K. Kronenberger, Continuous Mass Spectrometric Analysis of Fractions Separated by Gas Chromatography (in German), *Z. anal. Chem.* **189**:50–66 (1962).
33. Brunnée, C., L. Jenkel, and K. Kronenberger, Identification of Fractions which Have Been Separated in Gas Chromatographic Capillary Columns, ASTM E–14 Meeting on Mass Spectrometry, New Orleans, La., June 1962.
34. Brunnée, C., L. Jenkel, and K. Kronenberger, Advances in Use of Mass Spectrometers as Specific Ionization Detectors for Gas Chromatography (in German), *Z. anal. Chem.* **197**:42–50 (1963).
35. Butler, J. N., and R. D. McAlpine, Thermal *cis-trans* Isomerization of Crotononitrile, *Can. J. Chem.* **41**:2487–2491 (1963).
36. Butler, J. N., and G. J. Small, Thermal *cis-trans* Isomerization and Decomposition of Methyl Crotonate, *Can. J. Chem.* **41**:2492–2499 (1963).
37. Butler, J. N., and R. B. Ogawa, Thermal Decomposition of Cyclobutane at Low Pressures, *J. Am. Chem. Soc.* **85**:3346–3349 (1963).

38. Buttery, R. G., D. R. Black, M. P. Kealy, and W. H. McFadden, Volatile Hop Esters, *Nature* **202**:701–702 (1964).

39. Buttery, R. G., C. E. Hendel, and M. M. Boggs, Off-Flavors in Potato Products. Autoxidation of Potato Granules. Part I: Changes in Fatty Acids; Part II: Formation of Carbonyls and Hydrocarbons, *J. Agr. Food Chem.* **9**:245–248, 248–252 (1961).

40. Buttery, R. G., W. H. McFadden, R. Teranishi, M. P. Kealy, and T. R. Mon, Constituents of Hop Oil, *Nature* **200**:435–436 (1963).

41. Buttery, R. G., and R. Teranishi, Measurement of Fat Autoxidation and Browning Aldehydes in Food Vapors by Direct Vapor Injection Gas-Liquid Chromatography, *J. Agr. Food Chem.* **11**:504–507 (1963).

42. Camin, D. L., R. W. King, and S. D. Shawnan, Capillary Gas Chromatography Using Microvolume Thermal Conductivity Detectors, *Anal. Chem.* **36**:1175–1178 (1964).

43. Carnes, W. J., Composition of Straight Chain Alkylbenzenes by Gas Chromatography, *Anal. Chem.* **36**:1197–1200 (1964).

44. Caroti, G., New Developments in Solving Analytical Problems by Gas Chromatography (in Italian), *Chim. Ind. (Milan)* **45**:48–53 (1963).

45. Cartoni, G., A. Liberti, U. Pallotta, and R. Palombardi, Gas Chromatographic Determination of Elaidic Acid in Fats by Stereospecific Bromination (in Italian), *Riv. Ital. Sostanze Grasse* **40**:653–659 (1963).

46. Cartoni, G., A. Liberti, and G. Ruggieri, Gas Chromatographic and Spectrophotometric Study of the Isomerization Process of Polyunsaturated Fatty Acids (in Italian), *Riv. Ital. Sostanze Grasse* **40**:482–486 (1963).

47. Chovin, P., B. Thirion, and J. Tranchant, System for Introduction of Microvolumes of Samples in Gas Chromatography (in French), in: *Séparation Immédiate et Chromatographie,* GAMS, Paris, 1962, pp. 248–254.

48. Cieplinski, E. W., Separation of Cyclooctenols with Golay Columns, *Gas Chromatography Applications*, No. GC–DS–006, The Perkin–Elmer Corporation, Norwalk, Conn., 1963.

49. Cieplinski, E. W., Analysis of a Jet Fuel with Golay Column, *Gas Chromatography Applications*, No. GC–DS–012, The Perkin–Elmer Corporation, Norwalk, Conn., 1964.

50. Cieplinski, E. W., and W. Averill, Gas Chromatographic Analysis of Essential Oils Using Golay Columns and a Flame Ionization Detector, *Gas Chromatography Applications*, No. GC–AP–002, The Perkin–Elmer Corporation, Norwalk, Conn., 1963; also *Soap, Perfumery and Cosmetics Yearbook*, 1964, pp. 29–41.

51. Cieplinski, E. W., W. Averill, and F. J. Kabot, Analysis of Aromatic Hydrocarbons with Golay Columns, *Gas Chromatography Applications*, No. GC–DS–005, The Perkin–Elmer Corporation, Norwalk, Conn., 1963.

52. Cieplinski, E. W., L. S. Ettre, B. Kolb, and G. Kemmner, Pyrolysis–Gas Chromatography with Linearly Programmed Temperature Packed and Open Tubular Columns. The Thermal Degradation of Polyolefins, Part I, *Z. anal. Chem.* **205**:357–371 (1964).

53. Clarke, D. R., Quantitative Gas Stream Splitting Injection System Suitable for Use with Capillary Columns, *Nature* **198**:681–682 (1963).

54. Condon, R. D., Design Considerations of a Gas Chromatography System Employing High Efficiency Golay Columns, *Anal. Chem.* **31**:1717–1722 (1959).

55. Condon, R. D., Recent Advances in Golay Column Technology, Second Eastern Anal. Symposium, New York, Nov. 3, 1960.

56. Cremer, E., and M. Riedmann, Identification of Gas Chromatographically Separated Aromatic Substances in Honeys (in German), *Z. Naturforsch.* **19B**(1): 76–77 (1964).

57. Crowley, K. J., Photochemical Synthesis of β-Pinene, *Proc. Chem. Soc.* **1962**:245.

58. Davis, J. J., Compatible Readout System Designed for Use with a Chromatographic Instrument Employing Golay Columns and Ionization Detectors, in: *Gas Chromatography*, ed. H. J. Noebels, R. F. Wall, and N. Brenner, Academic Press, New York, 1961, pp. 85–90.

59. Davies, N. R., Palladium-Catalysed Olefin Isomerization, *Australian J. Chem.* **17**:212–218 (1964).

60. Del Bianco, F. M., Study on the Fatty Matter of the Cheese "Pecorino" (in Italian), *Riv. Ital. Sostanze Grasse* **40**:524–527 (1963).

61. De Jong, K., and H. Van der Wel, Identification of Some Iso-Linoleic Acids Occurring in Butterfat, *Nature* **202**:553–555 (1964).

62. Desty, D. H., Coated Capillary Columns, in: *Gas Chromatographie 1958*, ed. H. P. Angelé, Akademie Verlag, Berlin (Ost), 1959, pp. 176–184.

63. Desty, D. H., and A. Goldup, Coated Capillary Columns—An Investigation of Operating Conditions, in: *Gas Chromatography 1960*, ed. R. P. W. Scott, Butterworths, Washington, D.C., 1960, pp. 162–183.

64. Desty, D. H., A. Goldup, and W. T. Swanton, Separation of *m*-Xylene and *p*-Xylene by Gas Chromatography, *Nature* **183**:107–108 (1959).

65. Desty, D. H., A. Goldup, and W. T. Swanton, Performance of Coated Capillary Columns, in: *Gas Chromatography*, ed. N. Brenner, J. E. Callen, and M. D. Weiss, Academic Press, New York, 1962, pp. 105–135.

66. Desty, D. H., A. Goldup, and B. H. F. Whyman, Potentialities of Coated Capillary Columns for Gas Chromatography in the Petroleum Industry, *J. Inst. Petrol.* **45**:287–298 (1959).

67. Desty, D. H., J. N. Haresnape, and B. H. Whyman, Construction of Long Lengths of Coiled Glass Capillary, *Anal. Chem.* **32**:302–304 (1960).

68. Dijkstra, G., and J. De Goey, Use of Coated Capillaries as Columns for Gas Chromatography, in: *Gas Chromatography 1958*, ed. D. H. Desty, Butterworths, London, 1958, pp. 56–68.

69. Dorsey, J. A., R. H. Hunt, and M. J. O'Neal, Rapid Scanning Mass Spectrometry. Continuous Analysis of Fractions from Capillary Gas Chromatography, *Anal. Chem.* **35**:511–515 (1963).

70. Durrett, L. R., M. C. Simmons, and I. Dvoretzky, Quantitative Aspects of Capillary Gas Chromatography of Hydrocarbons, *Preprints*, Division of Petrol. Chem., Am. Chem. Soc. **6**(2–B):63–77 (1961).

71. Durrett, L. R., L. M. Taylor, C. F. Wantland, and I. Dvoretzky, Component Analysis of Isoparaffin-Olefin Alkylate by Capillary Gas Chromatography, *Preprints*, Division of Petrol. Chem., Am. Chem. Soc. **7**(3):5–13 (1962); also *Anal. Chem.* **35**:637–641 (1963).

72. Dvoretzky, I., D. B. Richardson, and L. R. Durrett, Applications of the Methylene Insertion Reaction to Component Analysis of Hydrocarbons, *Anal. Chem.* **35**:545–549 (1963).

73. Ettre, L. S., Quantitative Reliability of Hydrocarbon Analyses with a Capillary Column—Hydrogen Flame Ionization Detector System, Informal Symposium on Gas Chromatography, Section L of R.D.IV., ASTM D-2; 64th Annual ASTM Meeting, Atlantic City, N.J., June 27, 1961.

74. Ettre, L. S., Practical Aspects of Golay Column Gas Chromatography, *Instrument News* **13**(1a):1–8 (1961); also *Research & Development for Industry* **1962** (15):42–47.

75. Ettre, L. S., Wide Range Applications of Golay Columns, 13th Pittsburgh Conf. Anal. Chem. Appl. Spectroscopy, Pittsburgh, Pa., March 5, 1962.

76. Ettre, L. S., New Trends in Gas Chromatography and Their Applications, *Sci. Rept. Ist. Super. Sanita* **2**:252–272 (1962).

77. Ettre, L. S., Possibilities of Investigating and Expressing Column Efficiencies, *J. Gas Chromatog.* **1**(2):36–47 (1963).

78. Ettre, L. S., Open Tubular Columns in Gas Chromatographic Analysis, Fifth Eastern Anal. Symposium, New York, Nov. 14, 1963.

79. Ettre, L. S., New Developments in Column Technology in Gas Chromatography, ACHEMA—European Convention of Chemical Engineering, Frankfurt am Main, June 24, 1964.

80. Ettre, L. S., and W. Averill, Investigation of the Linearity of a Stream Splitter for Capillary Gas Chromatography, *Preprints*, Division of Petrol. Chem., Am. Chem. Soc. **6**(2–B):79–89 (1961); also *Anal. Chem.* **33**:680–684 (1961).

81. Ettre, L. S., and W. Averill, Recent Developments in Gas Chromatography with Respects to the Analysis of Very Small Quantities, in: *Proceedings—1961 International Symposium on Microchemical Techniques*, ed. N. D. Cheronis, Interscience, New York, 1962, pp. 715–732.

82. Ettre, L. S., W. Averill, and F. J. Kabot, Gas Chromatographic Analysis of Fatty Acids, *Gas Chromatography Applications*, No. GC–AP–001, The Perkin–Elmer Corporation, Norwalk, Conn., 1962.

83. Ettre, L. S., E. W. Cieplinski, and W. Averill, Application of Open Tubular (Golay) Columns with Larger Diameter, *J. Gas Chromatog.* **1**(2):7–16 (1963).

84. Ettre, L. S., E. W. Cieplinski, and N. Brenner, Quantitative Aspects of Capillary Gas Chromatography, *ISA Reprints*, No. 78–LA/61 (1961); also *ISA Transactions* **2**:134–140 (1963).

85. Ettre, L. S., E. W. Cieplinski, and F. J. Kabot, Analysis of Aromatics on Capillary Columns with Poly(Phenyl Ether) Type Liquid Phases, *J. Gas. Chromatog.* **1**(11): 38–40 (1963).

86. Ettre, L. S., V. J. Coates, and E. W. Cieplinski, New Developments in Gas Chromatographic Instrumentation and Their Application to Industrial Analysis (in German), in: *DECHEMA Monographien, Vol. 43*, ed. H. Bretschneider and K. Fischbeck, Verlag Chemie, Weinheim, 1962, pp. 241–255.

87. Ettre, L. S., R. D. Condon, F. J. Kabot, and E. W. Cieplinski, Dual Column—Differential Flame Ionization Detector System and Its Application with Both Packed and Open Tubular Columns, *Gas Chromatography Applications*, No. GC–AP–003, The Perkin–Elmer Corporation, Norwalk, Conn., 1963; also *J. Chromatog.* **13**:305–318 (1964).

88. Ettre, L. S., and F. J. Kabot, Quantitative Reproducibility of a Programmed Temperature Gas Chromatographic System with Constant Pressure Drop Using Packed and Golay Columns, *Anal. Chem.* **34**:1431–1434 (1962).

89. Ettre, L. S., and F. J. Kabot, Relative Response of Fatty Acid Methyl Esters on the Flame Ionization Detector, *J. Chromatog.* **11**:114–116 (1963).

90. Falconer, W. E., and R. J. Cvetanovic, Separation of Isotopically Substituted Hydrocarbons by Partition Chromatography, *Anal. Chem.* **34**:1064–1066 (1962).

91. Falconer, W. E., B. S. Rabinovitch, and R. J. Cvetanovic, Unimolecular Decomposition of Chemically Activated Propyl Radicals. Normal Intermolecular Secondary Kinetic Isotope Effect, *J. Chem. Phys.* **39**:40–53 (1963).

92. Farré-Rius, F., J. Henniker, and G. Guiochon, Wetting Phenomena in Gas Chromatography Capillary Columns, *Nature* **196**:63–64 (1962).

93. Felix, D., A. Melera, J. Seibl, and E. Kováts, Information on Etheric Oils. II. Structure of the So-called Linalooloxides (in German), *Helv. Chim. Acta* **46**: 1513–1536 (1963).

94. Fejes, P., J. Engelhardt, and G. Schay, A New Sampling System for Capillary Gas Chromatography (in German), in: *Gas Chromatographie 1963*, ed. H. P. Angelé and H. G. Struppe, Akademie Verlag, Berlin (Ost), 1963, pp. 21–31.

95. Fejes, P., J. Engelhardt, and G. Schay, New Sampling System for Capillary Chromatography (in German), *J. Chromatog.* **11**:151–157 (1963).

96. Fernandez, B., I. S. Fagerson, and W. W. Nawar, Applicability of Gas Chromatography to Detection of Changes in Orange Oil—A Preliminary Study, *J. Gas Chromatog.* **1**(9):21–22 (1963).

97. Fett, E. R., Backflush Applied to Capillary Column—Flame Ionization Detector Gas Chromatography System, *Anal. Chem.* **35**:419–420 (1963).

98. Freedman, R. W., and P. P. Croitoru, Quantitative Gas–Liquid Chromatography of Phenols by Complete Trimethylsilylation of Hindered Phenols in Presence of Acidic Oxides, *Anal. Chem.* **36**:1389–1390 (1964).

99. Freeman, S. K., Isomer Distribution of Some Chloromethylated Alkylbenzenes, *J. Org. Chem.* **26**:212–214 (1961).

100. Gast, L. E., W. L. Schneider, C. A. Forest, and J. C. Cowan, Composition of Methyl Esters from Heat-Bodied Linseed Oils, *J. Am. Oil Chemists' Soc.* **40**: 287–289 (1963).

101. Giacobbo, H., and W. Simon, Thermal Fragmentation and Structure Determination of Organic Substances, ACHEMA—European Convention of Chemical Engineering, Frankfurt am Main, June 24, 1964.

102. Giddings, J. C., Liquid Distribution on Gas Chromatographic Support. Relationship to Plate Height, *Anal. Chem.* **34**:458–465 (1962).

103. Giddings, J. C., Advances in the Theory of Plate Height in Gas Chromatography, *Anal. Chem.* **35**:439–449 (1963).

104. Giddings, J. C., Concepts and Column Parameters in Gas Chromatography, in: *Advances in Analytical Chemistry and Instrumentation*, ed. C. N. Reilley, Interscience—J. Wiley & Sons, New York, 1964, pp. 315–367.

105. Giddings, J. C., Theory of Gas Solid Chromatography. Potential for Analytical Use and the Study of Surface Kinetics, *Anal. Chem.* **36**:1170–1175 (1964).

106. Gill, H. A., and W. Averill, Design Considerations and Performance of a Linear Programmed Temperature Gas-Liquid Chromatograph for Golay and Packed Columns, 13th Pittsburgh Conf. Anal. Chem. Appl. Spectroscopy, Pittsburgh, Pa., March 5, 1962.

107. Gohlke, R. S., Time-of-Flight Mass Spectrometry: Application to Capillary Column Gas Chromatography, *Anal. Chem.* **34**:1332–1333 (1962).

108. Golay, M. J. E., Gas Chromatographic Terms and Definitions, *Nature* **182**: 1146–1147 (1958).

109. Golay, M. J. E., Theory and Practice of Gas Liquid Partition Chromatography with Coated Capillaries, in: *Gas Chromatography*, ed. V. J. Coates, H. J. Noebels, and I. S. Fagerson, Academic Press, New York, 1958, pp. 1–13.

110. Golay, M. J. E., Theory of Chromatography in Open and Coated Tubular Columns with Round and Rectangular Cross-Sections, in: *Gas Chromatography 1958*, ed. D. H. Desty, Butterworths, London, 1958, pp. 36–55.

111. Golay, M. J. E., Brief Report on Gas Chromatographic Theory, in: *Gas Chromatography 1960*, ed. R. P. W. Scott, Butterworths, Washington, D.C., 1960, pp. 139–143.

112. Golay, M. J. E. (assigned to The Perkin–Elmer Corp.), Vapor Fractometer Column, U.S. Patent 2,920,478 (1960).

113. Golay, M. J. E., Height Equivalent to a Theoretical Plate of Tubular Gas Chromatographic Columns Lined with a Porous Layer, *Nature* **199**:370–371 (1963).

114. Golay, M. J. E., L. S. Ettre and S. D. Norem, A Nomographic Approach to Some Problems in Linearly Programmed Temperature Gas Chromatography, in: *Gas Chromatography 1962*, ed. M. Van Swaay, Butterworths, Washington, D.C., 1962, pp. 139–151.

115. Grant, D. W., Potentialities of Capillary Columns in the Analysis of Coal Tar Fractions, *Coal Tar Res. Assoc. Report*, No. 0627, 1960.

116. Grant, D. W., Automatic Capillary Gas Chromatography and Sampling of Distillation Products, *Anal. Chem.* **36**:1519–1522 (1964).

117. Grant, D. W., and G. A. Vaughan, Analysis of Complex Phenolic Mixtures by Capillary Column Gas Liquid Chromatography after Silylation, in: *Gas Chromatography 1962*, ed. M. Van Swaay, Butterworths, Washington, D.C., 1962, pp. 305–314.

118. Griesbaum, K., A. A. Oswald, E. R. Quiram, and W. Naegele, Allene Chemistry, I. Free Radical Addition of Thiols to Allene, *J. Org. Chem.* **28**:1952–1957 (1963).

119. Grob, K., New Liquid Phases for the Gas Chromatographic Separation of Strong Bases on Capillary Columns, *J. Gas Chromatog.* **2**:80–82 (1964).

120. Habgood, H. W., and W. E. Harris, Capillary Programmed Temperature Gas Chromatography, *Anal. Chem.* **34**:882–885 (1962).

121. Hachenberg, H., and J. Gutberlet, Analytical Characterization of α-Olefin Mixtures with the Help of Programmed-Temperature Capillary Gas Chromatography, Infrared Spectroscopy, Bromine-Number Determination, and the FIA-Method (in German), *Brennstoff-Chem.* **45**:132–138 (1964).

122. Halász, I., E. Heine, C. Horváth, and H. G. Sternagel, Gas Chromatographic Analysis of C_1–C_7 Hydrocarbon Mixtures with "Solid Layer" and "Packed" Capillary Columns (in German), *Brennstoff-Chem.* **44**:387–389 (1963).

123. Halász, I., and C. Horváth, Thin-Layer Graphited Carbon Black as the Stationary Phase for Capillary Columns in Gas Chromatography, *Nature* **197**:71–72 (1963).

124. Halász, I., and C. Horváth, Open Tube Columns with Impregnated Thin Layer Support for Gas Chromatography, *Anal. Chem.* **35**:499–505 (1963).

125. Halász, I., and W. Schneider, Quantitative Gas Chromatographic Analysis of Hydrocarbons with Capillary Column and Flame Ionization Detector, *Anal. Chem.* **33**:979–982 (1961).

126. Halász, I., and W. Schneider, Quantitative Gas Chromatographic Analysis of Hydrocarbons with Capillary Columns and Flame Ionization Detector. II, in: *Gas Chromatography*, ed. N. Brenner, J. E. Callen, and M. D. Weiss, Academic Press, New York, 1962, pp. 287–306.

127. Halász, I., and G. Schreyer, Experiences with Capillary Column—Flame Ionization Detector Systems in Gas Chromatography (in German), *Chem.-Ing.-Tech.* **32**:675–685 (1960).

128. Halász, I., and G. Schreyer, Construction and Operation of a Capillary Column Chromatographic Apparatus with Flame Ionization Detector and Its Application in Quantitative Analysis (in German), *Z. anal. Chem.* **181**:384 (1961).

129. Halász, I., and G. Schreyer, Efficiency and Plate Number in Gas Chromatographic Analyses by Means of Capillary Columns (in German), *Z. anal. Chem.* **181**:367–382 (1961).

130. Hazeldean, G. S. F., and R. P. W. Scott, Resistance to Mass Transfer in Capillary Columns, *J. Inst. Petrol.* **48**:380–387 (1962).

131. Hayashi, S., C. Nakano, and T. Motoyama, Capillary Chromatography of Polyvinyl Acetate Emulsions (in Japanese), *Yukagaku* **12**:501–505 (1963).

132. Henneberg, D., and G. Schomburg, Mass Spectrometric Identification in Capillary Gas Chromatography, in: *Gas Chromatography 1962*, ed. M. Van Swaay, Butterworths, Washington, D.C., 1962, pp. 191–203.

133. Hennis, H. E., Diels-Alder Reactions. II. Independence of Reaction Temperature and Mode of Addition in the Isoprene-Methyl Acrylate Diels-Alder Reaction, *J. Org. Chem.* **28**:2570–2572 (1963).

134. Hill, D. W., J. R. Hook, and S. E. R. Mable, A Compact Cathode Ray Tube Gas Chromatograph, *J. Sci. Instr.* **39**:214–216 (1962).

135. Hill, D. W., and H. A. Newell, Effect of Nitrous Oxide and Carbon Dioxide on the Sensitivity of a "Small" Argon Detector for Use in Gas Chromatography, *Nature* **200**:1215–1216 (1964).

136. Hine, J., and H. E. Harris, Polar Effects on Rates and Equilibria. VII. Disproportionation and Isomerization of Alkyl- and Halobenzenes, *J. Am. Chem. Soc.* **85**:1476–1480 (1963).

137. Hively, R. A., Assignment of *cis-trans* Configuration to Mono-Olefin Pairs by Gas Chromatography, *Anal. Chem.* **35**:1921–1924 (1963).

138. Hollis, O. L., Gas–Liquid Chromatographic Analysis of Trace Impurities in Styrene Using Capillary Columns, *Anal. Chem.* **33**:352–355 (1961).

139. Hollis, O. L., and W. V. Hayes, Gas–Liquid Chromatographic Analysis of Chlorinated Hydrocarbons with Capillary Columns and Ionization Detectors, *Anal. Chem.* **34**:1223–1226 (1962).

140. Horton, A. D., Gas Chromatographic Analysis of Nitric Acid Degraded Di-*sec*-Butylphenyl-Phosphonate-Diethylbenzene, *AEC Rept.*, ORNL–3537, November 15, 1963, pp. 22–23.

141. Horváth, C., Separation Columns with Thin Porous Layers for Gas Chromatography (in German), Ph. D. Thesis, University of Frankfurt am Main, 1963.

142. Hückel, W., and R. Bross, Contribution to Constellation Analysis. X. Solvolysis of Esters of Toluene Sulfonic Acid. XIII. *cis-* and *trans*-3-Isopropyl-cyclopentanol (in German), *Ann. Chem.* **664**:1–19 (1963).

143. Hückel, W., and D. Rücker, Equilibria and Cleavage Reactions of Decalols and Monoalkylated Cyclohexanols (in German), *Ann. Chem.* **666**:30–44 (1963).

144. Hückel, W., and K. D. Thomas, Information on the Walden Inversion. X. *cis*-and *trans*-1-Methyl-2-amino-cyclopentane and Nitrous Acid (in German), *Chem. Ber.* **96**:2514–2518 (1963).

145. Hückel, W., and D. Volkman, Changes in the Molecular Structure through Chemical Reactions. XII. The 2,6-Shift in the Bicycloheptane Series (in German), *Ann. Chem.* **664**:31–75 (1963).

146. Hückel, W., and H. Waiblinger, Cleavage Velocity and Dipole Moments of 1,2-Dibromides (in German), *Ann. Chem.* **666**:17–30 (1963).

147. Hunter, G. L. K. and M. K. Veldhuis, A Temperature Programmed Preparative Scale Gas Chromatograph, *J. Chromatog.* **11**:11–18 (1963).

148. Hunter, G. L. K., A Simple Device for Cleaning and Coating Capillary Columns, *J. Chromatog.* **11**:124–125 (1963).

149. Jenkins, P., Reliable System for Introduction of Samples to Chromatographic Columns Particularly Suited for Quantitative Capillary Column Analysis, *Nature* **197**:72–73 (1963).

150. Jentzsch, D., New Gas Chromatographic Instruments (in German), in: *DECHEMA Monographien, Vol. 43*, ed. H. Bretschneider and K. Fischbeck, Verlag Chemie, Weinheim, 1962, pp. 231–240.

151. Jentzsch, D., and W. Hövermann, Possibilities of Application of Golay-Type Capillary Columns in Gas Chromatography (in German), *Aus der Praxis—für die Praxis*, No. 14–GC, Bodenseewerk Perkin–Elmer & Co., Überlingen, 1961.

152. Jentzsch, D., and W. Hövermann, Application of Macro Golay Columns in Gas Chromatography (in German), *Aus der Praxis—für die Praxis*, No. 19–GC, Bodenseewerk Perkin–Elmer & Co., Überlingen, 1962.

153. Jentzsch, D., and W. Hövermann, Critical Study on Golay Columns, in: *Gas Chromatography 1962*, ed. M. Van Swaay, Butterworths, Washington, D.C., 1962, pp. 204–215.

154. Jentzsch, D., and W. Hövermann, New Application of Golay Columns with Special Wall Treatments, 14th Pittsburgh Conf. Anal. Chem. Appl. Spectroscopy, Pittsburgh, Pa., March 4–8, 1963.

155. Jentzsch, D., and W. Hövermann, Application of Golay Columns of Larger Internal Diameter, *J. Chromatog.* **11**:440–451 (1963).

156. Johnstone, R. A. W., and P. M. Quan, Thermal Co-Dimerization of Isoprene and Buta-1,3-diene, *J. Chem. Soc. (London)* **1963**:935–938.

157. Kabot, F. J., W. Averill, and L. S. Ettre, Gas Chromatographic Analysis of Free Fatty Acids, *Riv. Ital. Sostanze Grasse* **41**:131–140 (1964).

158. Kabot, F. J., and L. S. Ettre, Gas Chromatography of Free Fatty Acids. III. Quantitative Aspects, *J. Gas Chromatog.* **1**(10):7–10 (1963).

159. Kabot, F. J., and L. S. Ettre, Analysis of Naphthalene Homologs by Gas Chromatography, *Anal. Chem.* **36**:250–252 (1964).

160. Kaiser, R., Chromatography in Gas Phase. Vol. II. *Capillary Gas Chromatography.* (Textbook; in German). Bibliographisches Institut, Mannheim, Germany, 1961. English translation, Butterworths, Washington, D.C., 1963.

161. Kaiser, R., New Results in the Application of Gas Chromatography (in German), *Z. anal. Chem.* **189**:1–14 (1962).

162. Kaiser, R. E., and H. G. Struppe, On the Theory and Experimental Conditions of Capillary Gas Chromatography (in German), in: *Gas Chromatographie 1959,*

ed. R. E. Kaiser and H. G. Struppe, Akademie Verlag, Berlin (Ost), 1959, pp. 177–194.

163 Kaiser, R., H. G. Struppe, H. Holzhäuser, and H. Kuhl, Gas Chromatographic Analysis of Gasoline (in German), *Freiberger Forschungsh.* **A192**:205–216 (1961); *C.A.* **56**(8):8992 (1962).

164. Karger, B. L., and W. D. Cooke, Effect of Column Length on Resolution under Normalized Time Conditions, *Anal. Chem.* **36**:985–991 (1964).

165. Kauffman, F. L., and G. D. Lee, Study of Octadecanoic Acids by Gas Liquid Partition Chromatography and Infrared Spectrophotometry, *J. Am. Oil Chemists' Soc.* **37**:385–386 (1960).

166. Kemmner, G., B. Kolb, and H. Pauschmann, Instrumentation, Technique and Application of Gas Chromatography in Mineral Oil Analysis, *Applied Gas Chromatography Series,* Bodenseewerk Perkin–Elmer & Co., Überlingen, 1963.

167. Keulemans, A. I. M., Progress in Gas Chromatography (in German), *Z. anal. Chem.* **170**:212–219 (1959).

168. Khan, M. A., Non-Equilibrium Theory of Capillary Columns and the Effect of Interfacial Resistance on Column Efficiency, in: *Gas Chromatography 1962,* ed. M. Van Swaay, Butterworths, Washington, D.C., 1962. pp. 3–17.

169. Kienitz, H., Modern Physical Methods of Analysis (in German), *Chem.-Ing.-Tech.* **32**:641–650 (1960).

170. Kirkland, J. J., Fibrillar Boehmite—A New Adsorbent for Gas Solid Chromatography, *Anal. Chem.* **35**:1295–1297 (1963).

171. Kiselev, A. V., The Importance of the Solid Surface in Partition, Adsorption, and Capillary Gas Chromatography, in: *Gas Chromatography 1962,* ed. M. Van Swaay, Butterworths, Washington, D.C., 1962, pp. xxxi–lii.

172. Knox, J. H., Speed of Analysis by Gas Chromatography, *J. Chem. Soc. (London)* **1961**:433–441.

173. Knox, J. H., and L. McLaren, The Spreading of Air Peaks in Capillary and Packed Gas Chromatographic Columns, *Anal. Chem.* **35**:449–454 (1963).

174. Kreyenbuhl, A., Construction of Pyrex Capillary Columns for Gas Chromatography (in French), *Bull. soc. chim. France* **1960**:2125–2127.

175. Landowne, R. A., and S. R. Lipsky, Use of Capillary Columns for the Separation of Some Closely Related Positional Isomers of Methyl Linoleate by Gas Chromatography, *Biochim. et Biophys. Acta* **46**:1–6 (1961).

176. Landowne, R. A., and S. R. Lipsky, Simple Method for Distinguishing between Unsaturated and Branched Fatty Acid Isomers by Gas Chromatography, *Biochim. et Biophys. Acta* **47**:589–592 (1961).

177. Leibnitz, W., and M. Mohnke, Surface Modified Glass Capillaries for Polar Separation Fluids in Gas Chromatography (in German), *Chem. Tech. (Berlin)* **14**:753–754 (1962).

178. Le Tourneau, R. L., Analysis of Petroleum, *Preprints*, Division of Petrol. Chem., Am. Chem. Soc. **6**(3–A):15–25 (1961).

179. Liberti, A., G. P. Cartoni, and F. Bruner, Isotopic Effects in Gas Liquid Chromatography. I. Gas Chromatographic Behaviour of Benzene and Perdeuterobenzene, *J. Chromatog.* **12**:8–14 (1963).

180. Lipsky, S. R., and R. A. Landowne, Effects of Varying the Chemical Composition of the Stationary Phase on the Separation of Certain C_{19}, C_{21}, and C_{27} Steroids by Gas Chromatography, *Anal. Chem.* **33**:818–828 (1961).

181. Lipsky, S. R., R. A. Landowne, and J. E. Lovelock, Separation of Lipids by Gas Chromatography, *Anal. Chem.* **31**:852–856 (1959).

182. Lipsky, S. R., J. E. Lovelock, and R. A. Landowne, Use of High Efficiency Capillary Columns for the Separation of Certain *cis-trans* Isomers of Long Chain Fatty Acid Esters by Gas Chromatography, *J. Am. Chem. Soc.* **81**:1010 (1959).

183. Litchfield, C., A. F. Isbell, and R. Reiser, Analysis of the Geometric Isomers of Methyl Linoleate by Gas Chromatography, *J. Am. Oil Chemists' Soc.* **39**:330–334 (1962).

184. Litchfield, C., J. E. Lord, A. F. Isbell, and R. Reiser, *Cis-trans* Isomerization of Oleic, Linoleic, and Linolenic Acids, *J. Am. Oil Chemists' Soc.* **40**:553–557 (1963).

185. Litchfield, C., R. Reiser, and A. A. Isbell, Analysis of *cis-trans* Fatty Acid Isomers Using Gas–Liquid Chromatography, *J. Am. Oil Chemists' Soc.* **40**:302–309 (1963).

186. Litchfield, C., R. Reiser, A. Isbell, and G. L. Feldman, Gas Chromatography of *cis-trans* Fatty Acid Isomers on Nitrile Silicone Capillary Columns, *J. Am. Oil Chemists' Soc.* **41**:52–55 (1964).

187. Lovelock, J. E., Detector for Use with Capillary Tube Columns in Gas Chromatography, *Nature* **182**:1663–1664 (1958).

187/a. Lovelock, J. E., and N. L. Gregory (with the additional remark of A. Zlatkis), Electron Capture Ionization Detector, in: *Gas Chromatography*, ed. N. Brenner, J. E. Callen, and M. D. Weiss, Academic Press, New York, 1962, pp. 219–229.

188. Mackay, D. A. M., Coming—An Analytical Revolution, *Food Eng.* **31**(7):82–83 (1959).

189. Mair, B. J., and T. J. Mayer, Composition of the Dinuclear Aromatics, C_{12} to C_{14} in the Light Gas Oil Fraction of Petroleum, *Preprints*, Division of Petrol. Chem., Am. Chem. Soc. **8**(4–C):29–51 (1963).

190. Marco, J. R. P., High Speed Analysis by Capillary Column—Programmed Temperature Gas Chromatography, *GC Newsletter* (The Perkin–Elmer Corp.) **1**(3):1–2 (1964).

191. Martin, R. L., and J. C. Winters, Determination of Hydrocarbons in Crude Oil by Capillary-Column Gas Chromatography, *Preprints*, Division of Petrol. Chem., Am. Chem. Soc. **8**(4–C):67–75 (1963); also *Anal. Chem.* **35**:1930–1933 (1963).

192. Martin, R. L., J. C. Winters, and J. A. Williams, Composition of Crude Oils by Gas Chromatography: Geological Significance of Hydrocarbon Distribution, *Preprints*, No. V/13, Fifth World Petrol. Congress, Frankfurt am Main, June 19–26, 1963.

193. McBride, J. J., E. Jungermann, J. U. Killheffer, and R. J. Clutter, New Phosphorylation Reaction of Olefins. II. A Novel Synthesis of a Four-Membered Phosphorous Containing Ring Compound, *J. Org. Chem.* **27**:1833–1836 (1962).

194. McEwen, D. J., Improved Sampling Valve for Gas Chromatography, *J. Chromatog.* **9**:266–269 (1962).

195. McEwen, D. J., Temperature Programmed Capillary Columns in Gas Chromatography, *Anal. Chem.* **35**:1636–1640 (1963).

196. McEwen, D. J., Backflush and Two-Stage Operation of Capillary Columns in Gas Chromatography, *Anal. Chem.* **36**:279–282 (1964).

197. McFadden, W., Mass Spectrometry with Capillary Gas Chromatography, *Chemistry in Canada* **15**(4):54 (1963).

198. McFadden, W. H., and R. Teranishi, Fast Scan Mass Spectrometry with Capillary

Gas–Liquid Chromatography in Investigation of Fruit Volatiles, *Nature* **200**: 329–330 (1963).

199. McFadden, W. H., R. Teranishi, D. R. Black, and J. C. Day, Use of Capillary Gas Chromatography with Time-of-Flight Mass Spectrometer, *J. Food Sci.* **28**:316–319 (1963).

200. Mohnke, M., and W. Saffert, Adsorption Chromatography of Hydrogen Isotopes with Capillary Columns, in: *Gas Chromatography 1962*, ed. M. Van Swaay, Butterworths, Washington, D.C., 1962, pp. 216–224.

201. Napier, D. H., and J. R. Simonson, Metering Valve for Capillary Chromatography Samples, *Chem. Ind. (London)* **1962**(42):1831–1832.

202. Navar, W. W., and I. S. Fagerson, Direct Gas Chromatographic Analysis as an Objective Method of Flavor Measurement, *Food Technol.* **16**:107–109 (Nov. 1962).

203. Negri, R. C., Developments in Gas Chromatography (in Spanish), *Afinidad* **18**: 383–394 (1961).

204. Norem, S. R., Behavior of Inert Gas Packets in Chromatographic Columns, *Anal. Chem.* **34**:40–42 (1962).

205. Nordström, K., Formation of Esters from Acids by Brewer's Yeast. II. Formation from Lower Fatty Acids, *J. Inst. Brewing* **70**(1):42–55 (1964).

206. Olah, G. A., S. J. Kuhn, S. H. Flood, and B. A. Hardie, Aromatic Substitution. XIV. Ferric Chloride Catalyzed Bromination of Benzene and Alkylbenzenes with Bromine in Nitromethane Solution, *J. Am. Chem. Soc.* **86**:1039–1044 (1964).

207. Olah, G. A., S. J. Kuhn, S. H. Flood, and B. A. Hardie, Aromatic Substitution. XV. Ferric Chloride Catalyzed Bromination of Halobenzenes in Nitromethane Solution, *J. Am. Chem. Soc.* **86**:1044–1046 (1964).

208. Olah, G. A., S. H. Flood, S. J. Kuhn, M. E. Moffatt, and N. A. Overchuck, Aromatic Substitution. XVI. Friedel–Crafts Isopropylation of Benzene and Methylbenzenes with Isopropyl Bromide and Propylene, *J. Am. Chem. Soc.* **86**:1046–1054 (1964).

209. Olah, G. A., S. J. Kuhn, and B. A. Hardie, Aromatic Substitution. XVII. Ferric Chloride and Aluminum Chloride Catalyzed Chlorination of Benzene, Alkylbenzenes and Halobenzenes, *J. Am. Chem. Soc.* **86**:1055–1060 (1964).

210. Olah, G. A., S. H. Flood, and M. E. Moffatt, Aromatic Substitution. XIX. Friedel–Crafts Isopropylation and *t*-Butylation of Halobenzenes, *J. Am. Chem. Soc.* **86**:1065–1066 (1964).

211. Oswald, A. A., Griesbaum, K., and B. E. Hudson, Organic Sulfur Compounds. Amine-Hydroperoxide Complexes as Intermediates in the Cooxidation of Thiols with 2,5-Dimethyl-2,4-hexadiene, *J. Org. Chem.* **28**:2351–2354 (1963).

212. Oyama, V. I., Use of Gas Chromatography for the Detection of Life on Mars, *Nature* **200**:1058–1059 (1963).

213. Pallotta, U., G. P. Cartoni, and A. Liberti, Critical Evaluation of Some Vegetable Oil Modifications in Industrial Processes (in Italian), *Riv. Ital. Sostanze Grasse* **40**:487–493 (1963).

214. Patton, S., R. D. McCarthy, L. E. Evans, and T. R. Lynn, Structure and Synthesis of Milk Fat. I. Gas Chromatographic Analysis, *J. Dairy Sci.* **43**:1187–1195 (1960).

215. Perkin–Elmer Corp., Separation Column for Gas Chromatographic Instruments and Instrument for the Application of Such Columns (in German), German Patent 1,063,409 (1959).

216. Perkin–Elmer Corp., Gas Chromatographic Columns, Brit. Patent 834, 390 (1960).

217. Petitjean, D. L., and C. J. Leftault, Jr., Oxide-Coated Aluminum Tubing for Capillary Gas Chromatography, *J. Gas Chromatog.* **1**(3):18–21 (1963).

218. Petrocelli, J. A., Modified Thermal Conductivity Detector for Capillary Columns, *Anal. Chem.* **35**:2220–2221 (1963).

219. Phillips, T. R., and D. R. Owens, Gas Chromatographic Analysis of Inorganic Halogen Compounds on Capillary Columns, in: *Gas Chromatography 1960*, ed. R. P. W. Scott, Butterworths, Washington, D.C., 1962, pp. 308–317.

220. Pietropaolo, C., G. Cali, L. Pisano, and E. Fiorentino, Studies on the Fatty Acid Composition of Pleural and Peritoneal Exudates (in Italian), *Bull. Soc. Ital. Biol. Sper.* **39**:1601–1602 (1963).

221. Polgár, A. G., J. J. Holst, and S. Groenning, Determination of Alkanes and Cycloalkanes through C_8 and Alkanes through C_7 by Capillary Gas Chromatography, *Anal. Chem.* **34**:1226–1234 (1962).

222. Porcaro, P. J., Observations on the Use of "Empty" Copper Tubular Capillary Columns, *J. Gas Chromatog.* **1**(6):17–19 (1963).

223. Prévot, A., Newest Applications of Gas Chromatography in the Analysis of Lipids (in French), *Bull. soc. chim. France* **1963**:314–316.

224. Purcell, J. E., Separation of Inert Gases on Open Tubular Adsorption Columns, *Nature* **201**:1321 (1964).

225. Purnell, J. H., Comparison of Efficiency and Separating Power of Packed and Capillary Gas Chromatographic Columns, *Nature* **184**:2009 (1959).

226. Purnell, J. H., Correlation of Separating Power and Efficiency of Gas Chromatographic Columns, *J. Chem. Soc. (London)* **1960**:1268–1274.

227. Purnell, J. H., and C. P. Quinn, An Approach to Higher Speeds in Gas–Liquid Chromatography, in: *Gas Chromatography 1960*, ed. R. P. W. Scott, Butterworths, Washington, D.C., 1960, pp. 184–198.

228. Quiram, E. R., Applications of Wide-Diameter Open Tubular Columns in Gas Chromatography, *Anal. Chem.* **35**:593–595 (1963).

229. Richardson, D. B., M. C. Simmons, and I. Dvoretzky, Reactivity of Methylene from Photolysis of Diazomethane, *J. Am. Chem. Soc.* **83**:1934–1947 (1961).

230. Rödel, E., Working Methods of Gas Chromatography, ACHEMA—European Convention of Chemical Engineering, Frankfurt am Main, June 24, 1964.

231. Rouayheb, G. M., O. F. Folmer, and W. C. Hamilton, Quantitative Gas Chromatographic Analysis of Hydrocarbon Systems Using the Lovelock Diode Detector and Capillary Columns, *Anal. Chim. Acta* **26**:378–390 (1962).

232. Rouayheb, G. M., and W. C. Hamilton, Deterioration of Solid Coated Capillary Columns on Standing, *Nature* **191**:801–802 (1961).

233. Sasaki, N., K. Tominaga, and M. Aoyagi, Micro Gas Chromatograph, *Nature* **186**:309–310 (1960).

234. Schneck, E., Importance of Gas Chromatography for the Development and Plant Control of a High-Temperature Cracking Process (in German), *Brennstoff-Chem.* **44**(11):354–360 (1963).

235. Schneider, R. A., J. P. Costiloe, A. Vega, and S. Wolf, Olfactory Threshold Technique with Nitrogen Dilution of *n*-Butane and Gas Chromatography, *J. Appl. Physiol.* **18**:414–417 (1963).

236. Scholfield, R. C., E. P. Jones, R. O. Butterfield, and H. J. Dutton, Argentation

in Countercurrent Distribution to Separate Isologues and Geometric Isomers of Fatty Acid Esters, *Anal. Chem.* **35**:1588–1591 (1963).

237. Scholfield, C. R., E. P. Jones, J. Nowakowska, E. Selke, B. Sreenivasan, and H. J. Dutton, Hydrogenation of Linoleate. I. Fractionation and Characterization Studies, *J. Am. Oil Chemists' Soc.* **37**:579–582 (1960).

238. Scholfield, C. R., E. P. Jones, J. Nowakowska, and H. J. Dutton, Hydrogenation of Linolenate. II. Hydrazine Reduction, *J. Am. Oil Chemists' Soc.* **38**:208–211 (1961).

239. Schreyer, G., Experiences with Capillary Columns and Flame Ionization Detectors in Gas Chromatography, Ph. D. Thesis, University of Frankfurt am Main, 1961.

240. Schwartz, R. D., and D. J. Brasseaux, Resolution of Complex Hydrocarbon Mixtures by Capillary Column Gas Liquid Chromatography—Composition of the 28–114°C Portion of Petroleum, *Preprints*, Division of Petrol. Chem., Am. Chem. Soc. **8**(4–C):77–86 (1963); also *Anal. Chem.* **35**:1374–1382 (1963).

241. Schwartz, R. D., D. J. Brasseaux, and G. R. Shoemake, Capillary Column Gas–Liquid Chromatography with Thermal Conductivity Detectors, *J. Gas Chromatog.* **1**(1):32–33; (2):17 (1963).

242. Schwartz, R. D., D. J. Brasseaux, and G. R. Shoemake, Sol-Coated Capillary Adsorption Columns for Gas Chromatography, *Anal. Chem.* **35**:496–499 (1963).

243. Scott, R. P. W., Nylon Capillary Columns for Use in Gas Chromatography, *Nature* **183**:1753–1754 (1959).

244. Scott, R. P. W., Cathode Ray Presentation of Chromatograms, *Nature* **185**: 312–313 (1960).

245. Scott, R. P. W., An Introductory Lecture to Apparatus and Technique, in: *Gas Chromatography 1960*, ed. R. P. W. Scott, Butterworths, Washington, D.C., 1960, pp. 3–6.

246. Scott, R. P. W., Process Monitoring by Gas Chromatography, *Research Applied on Industry* **14**:113–117 (1961).

247. Scott, R. P. W., Effect of Temperature on the Efficiency, Resolution and Analysis Time of Capillary Columns, *J. Inst. Petrol.* **47**:284–290 (1961).

248. Scott, R. P. W., and C. A. Cumming, Cathode Ray Presentation of Chromatograms, in: *Gas Chromatography 1960*, ed. R. P. W. Scott, Butterworths, Washington, D.C., 1960, pp. 117–128.

249. Scott, R. P. W., and G. S. F. Hazeldean, Some Factors Affecting Column Efficiency and Resolution of Nylon Capillary Columns, in: *Gas Chromatography 1960*, ed. R. P. W. Scott, Butterworths, Washington, D.C., 1960, pp. 144–161.

250. Self, R., An Enrichment Trap for Use with Capillary Columns, *Nature* **189**:223 (1961).

251. Self, R., J. C. Casey, and T. Swain, Low Boiling Volatiles of Cooked Foods, *Chem. Ind. (London)* **1963**:863–864.

252. Self, R., D. G. Land, and J. C. Casey, Gas Chromatography Using Capillary Column Units for Flavour Investigation, *J. Sci. Food Agr.* **14**:209–220 (1963).

253. Self, R., and T. Swain, Flavour in Potatoes, *Proc. Nutrition Soc.* **22**:176–182 (1963).

254. Simmons, M. C., D. B. Richardson, and I. Dvoretzky, Structural Analysis of Hydrocarbons by Capillary Gas Chromatography in Conjunction with the Methylene Insertion Reaction, in: *Gas Chromatography 1960*, ed. R. P. W. Scott, Butterworths, Washington, D.C., 1960, pp. 211–223.

255. Simon, M. S., Spectral Shifts in Anthraquinone Dyes Caused by Non-Conjugated Substituents, *J. Am. Oil Chemists' Soc.* **85**:1969–1974 (1963).

256. Smith, B., and M. Erwik, Determination of Hydrocarbon Constituents in the Benzene Pre-Run, *Acta Chim. Scand.* **17**:283–295 (1963).

256/a. Smith, G. A. L., and D. A. King, Separation and Identification of Steam Volatile Phenols Present in Cigarette Smoke Condensate by Capillary Column Gas Liquid Chromatography, *Chem. Ind. (London)* (13):540–541 (1964).

257. Sreenivasan, B., J. Nowakowska, E. P. Jones, E. Selke, C. R. Scholfield, and H. J. Dutton, Hydrogenation of Linolenate. VII. Separation and Identification of Isomeric Dienes and Monoenes, *J. Am. Oil Chemists' Soc.* **40**:45–51 (1963).

258. Sternberg, J. C., and R. E. Poulson, Particle-to-Column Diameter Ratio. Effect on Band Spreading, *Anal. Chem.* **36**:1492–1502 (1964).

259. Strickler, H., and E. Kováts, Influence of Experimental Conditions on Peak Resolution in Gas Chromatography, *J. Chromatog.* **8**:289–302 (1962).

260. Struppe, H. G., Evaluation of Capillary Columns, in: *Gas Chromatographie 1961*, ed. H. P. Angelé and H. G. Struppe, Akademie Verlag, Berlin (Ost), 1961, pp. 250–274.

261. Struppe, H. G., Reaction—Gas Chromatography on Capillary Columns at the Hydrogenation of Olefins (in German), *Chem. Tech. (Berlin)* **14**:114 (1962).

262. Struppe, H. G., The Error in Determination of Retention Data and Its Influence on the Identification in Capillary Gas Chromatography (in German), in: *Gas Chromatographie 1963*, ed. H. P. Angelé and H. G. Struppe, Akademie Verlag, Berlin (Ost), 1963, pp. 378–401.

263. Swartz, D. J., K. W. Wilson, and W. J. King, Merits of Liquefied Petroleum Gas Fuel for Automotive Air Pollution Abatement, *J. Air Pollution Control Assoc.* **13**(4):154–159 (1963).

264. Teranishi, R., and R. G. Buttery, Aromagrams—Direct Vapor Analyses with Gas Chromatography, in: *Reports of the Scientific Technical Commission, International Federation of Fruit Juice Producers*, Juris Verlag, Zurich, 1962, pp. 257–266.

265. Teranishi, R., R. G. Buttery, R. E. Lundin, W. H. McFadden, and T. R. Mon, Role of Gas Chromatography in Aroma Research, in: *Am. Soc. of Brewing Chemists, Proceedings 1963*, pp. 52–57.

266. Teranishi, R., R. G. Buttery, W. H. McFadden, T. R. Mon, and J. Wasserman, Capillary Column Efficiencies in Gas Chromatography—Mass Spectral Analyses, *Anal. Chem.* **36**:1509–1512 (1964).

267. Teranishi, R., J. W. Corse, W. H. McFadden, D. R. Black, and A. I. Morgan, Jr., Volatiles from Strawberries. I. Mass Spectral Identification of the More Volatile Components, *J. Food Sci.* **28**:478–483 (1963).

268. Teranishi, R., and T. R. Mon, Large-Bore Capillary and Low Pressure Drop Packed Columns, *Anal. Chem.* **36**:1491–1492 (1964).

269. Teranishi, R. C., C. C. Nimmo, and J. Corse, Programmed Temperature Control of Capillary Column, *Anal. Chem.* **32**:1384–1386 (1960).

270. Teranishi, R., T. H. Schultz, W. H. McFadden, R. E. Lundin, and D. R. Black, Volatiles from Oranges. I. Hydrocarbons Identified by Infrared, Nuclear Magnetic Resonance and Mass Spectra, *J. Food Sci.* **28**:541–545 (1963).

271. Thijssen, H. A. C., Gas–Liquid Chromatography—A Contribution to the Theory of Separation in Open Hole Tubes, *J. Chromatog.* **11**:141–150 (1963).

272. Tobe, B. A., Metabolism of Volatile Amines. VI. Identification of the Volatile Base Present in Blood, *Can. Med. Assoc. J.* **89**:1320–1324 (1963).
273. Varadi, P. F., and K. Ettre, Vacuum Output Gas Chromatography, *Anal. Chem.* **35**:410–412 (1963).
274. Walker, J. Q., Better Efficiency, Greater Detection Sensitivity, Speed in Analysis, *Oil & Gas J.* Apr. 8, 1963, pp. 78–80.
275. Walker, J. Q., Separation of Phenol and Substituted Methyl Phenols, *J. Gas Chromatog.* **2**:46 (1964).
276. Walker, J. Q., and D. L. Ahlberg, Quantitative Analysis of Aromatic Hydrocarbons by Capillary Gas Chromatography, *Anal. Chem.* **35**:2022–2027 (1963).
277. Walker, J. Q., and D. L. Ahlberg, Qualitative Analysis of Naphthenes by Capillary Gas Chromatography, *Anal. Chem.* **35**:2028–2030 (1963).
278. Walz, H., Investigation of an Isomeric Mixture with Preparative Gas Chromatography and Quantitative Determination with Golay Column—Flame Ionization Detector (in German), *Aus der Praxis—für die Praxis*, No. 15–GC, Bodenseewerk Perkin–Elmer & Co., Überlingen, 1962.
279. Weingarten, H., W. D. Ross, J. M. Schlater, and G. Wheeler, Jr., Gas Chromatographic Analysis of Chlorinated Biphenyls, *Anal. Chim. Acta* **26**:391–394 (1962).
280. Wiseman, W. A., Separation Factors in Gas Chromatography, *Nature* **185**: 841–842 (1960).
281. Zimmerman, C. A., J. T. Kelly, and J. C. Dean, Dimethylhexane Formation in Butene Alkylation, *Ind. Eng. Chem. Products Res. & Development* **1**:124–126 (1962).
282. Zlatkis, A., Ionization Detectors and Capillary Columns, in: *Lectures on Gas Chromatography 1962*, ed. H. A. Szymanski, Plenum Press, New York, 1963, pp. 87–104.
283. Zlatkis, A., and H. R. Kaufmann, Use of Coated Tubing for Gas Chromatography, *Nature* **184**:2010 (1959).
284. Zlatkis, A., and J. E. Lovelock, Gas Chromatography of Hydrocarbons Using Capillary Columns and Ionization Detectors, *Anal. Chem.* **31**:620–621 (1959).
285. Zlatkis, A., and J. Q. Walker, Surface Modification of Capillary Columns for Use in Gas Chromatography, *Anal. Chem.* **35**:1359–1362 (1963).
286. Zlatkis, A., and J. Q. Walker, Direct Sample Introduction for Large Bore Capillary Columns in Gas Chromatography. *J. Gas Chromatog.* **1**(5):9–11 (1963).

SUBJECT INDEX TO THE BIBLIOGRAPHY

5. Applications

Kinetic and catalytic studies 1, 16, 17, 23, 57, 59, 91, 136, 137

Study of organic reactions 16, 17, 26, 35, 36, 37, 133, 140, 142, 143, 144, 145, 146, 156, 193, 229

Pyrolysis 52, 101, 131

Quantitative analysis 25, 28, 70, 73, 80, 84, 88, 98, 125, 126, 128, 231, 278

Hydrogenation 261

Process monitoring 116, 234, 246

High-speed analysis 65, 227, 244, 248

Applied chemistry

  Biochemistry 22, 212, 220, 272

  Food industry 18, 38, 39, 40, 41, 45, 56, 60, 61, 82, 198, 205, 213, 214, 223, 250, 251, 252, 253

  Essential oils 7, 9, 13, 20, 21, 50, 93, 96, 188, 198, 202, 252, 253, 264, 265, 270

  Petroleum industry 2, 15, 49, 66, 70, 71, 72, 116, 163, 166, 189, 191, 192, 240, 254, 256, 263, 267

  Isotopes 29, 90, 177, 179, 200

  Tar products 12, 98, 115, 117

Sixth Part

Supplements

6.1 CALCULATION OF THE "AIR PEAK" TIME

As mentioned in chapter 3.32, the calculation of the "air peak" time utilizes the known fact that a straight-line relationship exists between the logarithm of the adjusted retention time and the number of carbon atoms in a homologous series:

$$n_c = m \log t_R' + q = m \log (t_R - t_M) + q \tag{64}$$

where n_c is the chain length (carbon number) of the homolog, t_R is its retention time (measured from the point of injection), t_R' is the adjusted retention time (measured from the "air peak"), t_M is the retention time of an inert substance (usually air), m and q are constants.

Two methods are described in the literature for solving Eq. (64).

6.11 The method of Peterson and Hirsch[51]

For this calculation, the chromatogram must contain the peaks of three members of a homologous series whose carbon numbers (n_c) fulfill the following condition:

$$n_{c2} - n_{c1} = n_{c3} - n_{c2} \tag{65}$$

Figure 77 illustrates the method of calculation. An arbitrary line (line "O") is drawn on the chromatogram and the distances of the peak maxima from this line are called x_1, x_2, and x_3, respectively. The distance of line "O" from the "air peak" maximum is x_0; thus, the adjusted retention times can be expressed as $(x_1 + x_0)$, $(x_2 + x_0)$, and $(x_3 + x_0)$. These values can be substituted into Eq. (64):

$$n_{c1} = m \log(x_1 + x_0) + q \tag{66a}$$

$$n_{c2} = m \log(x_2 + x_0) + q \tag{66b}$$

$$n_{c3} = m \log(x_3 + x_0) + q \tag{66c}$$

Subtracting Eq. (66a) from Eq. (66b) and Eq. (66b) from Eq. (66c), we obtain

$$n_{c2} - n_{c1} = m \log \frac{x_2 + x_0}{x_1 + x_0} \tag{67a}$$

[51] M. L. Peterson and J. Hirsch, *J. Lipid Res.* **1**:132 (1959).

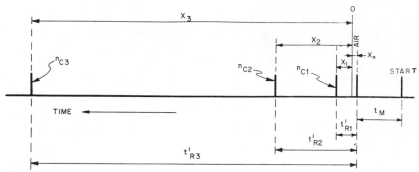

Figure 77. Calculation of the air peak time using the method of Peterson and Hirsch. For the sake of simplicity, the individual peaks are represented only by a vertical line drawn through their maxima.

$$n_{c3} - n_{c2} = m \log \frac{x_3 + x_0}{x_2 + x_0} \tag{67b}$$

Substituting the right-hand sides of these equations into Eq. (65), we get

$$\frac{x_2 + x_0}{x_1 + x_0} = \frac{x_3 + x_0}{x_2 + x_0} \tag{68}$$

or

$$x_0 = \frac{x_2^2 - x_1 x_3}{x_1 + x_3 - 2x_2} \tag{69}$$

Example 1.

Let us suppose that on the original chromatogram corresponding to Figure 44, the "O" line is drawn 33.5 mm from the start, and that the peaks corresponding to the C_4, C_6, and C_8 acids are selected to satisfy Eq. (65). The distances of the peak maxima from the "O" line then are (chart speed: 0.5 inch/min):

$$x_1 = 10 \text{ mm}$$
$$x_2 = 50 \text{ mm}$$
$$x_3 = 209 \text{ mm}$$

Substitution of these values into Eq. (69) yields $x_0 = 3.5$ mm. Thus,

$$t_M = 33.5 - 3.5 = 30 \text{ mm}$$

and the corresponding adjusted retention times are

$$t'_{R1} = 10 + 3.5 = 13.5 \text{ mm}$$
$$t'_{R2} = 50 + 3.5 = 53.5 \text{ mm}$$
$$t'_{R3} = 209 + 3.5 = 212.5 \text{ mm}$$

If the value of x_0 calculated from Eq. (69) is positive, the air peak maximum precedes the "O" line; if x_0 is negative, it follows the "O" line.

This calculation can be further simplified by designating the peak maximum of the second homolog as the "O" line (see Figure 78). In this case, x_2 will be zero and

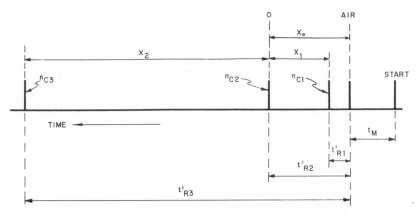

Figure 78. Calculation of the air peak time using the simplified method of Peterson and Hirsch. For the sake of simplicity, the individual peaks are represented only by a vertical line drawn through their maxima.

suitable modification of Eq. (69) results in the following equation for x_0, i.e., the adjusted retention time of the second homolog (x_1 is now a negative number):

$$x_0 = \frac{x_1 x_3}{x_3 - x_1} \tag{70}$$

Example 2.

We use the same chromatogram as for Example 1; the values now are

$$x_1 = 40 \text{ mm}$$
$$x_2 = 159 \text{ mm}$$

Substituting these values into Eq. (70) yields

$$x_0 = t'_{R2} = 53.5 \text{ mm}$$

and the two other adjusted retention times are

$$t'_{R1} = 13.5 \text{ mm}$$
$$t'_{R2} = 212.5 \text{ mm}$$

Example 3.

A further example for this calculation is the analysis of a hydrocarbon mixture on a 300 ft × 1.55 mm ID open tubular column coated with squalane liquid phase, at 60°C, and a carrier gas (He) flow of 13.2 ml/min. Three of the components belonged to one homologous series, with the following retention times:

Component	Carbon number	t_R (mm)
n-pentane	5	112
n-hexane	6	127
n-heptane	7	166

If the peak maximum of the second component is taken as the "O" line, the proper values used in Eq. (70) are

$$x_1 = 127 - 112 = 15 \text{ mm}$$
$$x_3 = 166 - 127 = 39 \text{ mm}$$

Substituting these values into Eq. (70) yields

$$x_0 = t'_{R2} = 24.4 \text{ mm}$$

Thus,

$$t_M = 127 - 24.4 = 102.6 \text{ mm}$$

This actual analysis was carried out with a thermistor detector and, thus, the air peak was also observed, with a retention time of 102 mm. As can be seen, the agreement between the calculated and measured values is good.

6.12 The Method of Gold[52]

The only restriction of the method of Peterson and Hirsch is that the three homologs used for the calculation have to be evenly spaced with respect to carbon number. The method of Gold, on the other hand, permits the use of any three homologs.

For this calculation, let us start with Eq. (64) and write it for three homologs with carbon numbers n_{c1}, n_{c2}, and n_{c3}, respectively:

$$n_{c1} = m \log(t_{R1} - t_M) + q \tag{71a}$$

$$n_{c2} = m \log(t_{R2} - t_M) + q \tag{71b}$$

$$n_{c3} = m \log(t_{R3} - t_M) + q \tag{71c}$$

Subtracting Eq. (71a) from Eq. (71b) and Eq. (71c), respectively, we obtain

$$n_{c2} - n_{c1} = m \log \frac{t_{R2} - t_M}{t_{R1} - t_M} \tag{72a}$$

$$n_{c3} - n_{c1} = m \log \frac{t_{R3} - t_M}{t_{R1} - t_M} \tag{72b}$$

We now solve these equations for m:

$$m = \frac{n_{c2} - n_{c1}}{\log \dfrac{t_{R2} - t_M}{t_{R1} - t_M}} \tag{73a}$$

$$m = \frac{n_{c3} - n_{c1}}{\log \dfrac{t_{R3} - t_M}{t_{R1} - t_M}} \tag{73b}$$

For these equations, the retention times (i.e., the distances between the start and the three peak maxima) can be obtained from the chromatogram; $(n_{c2} - n_{c1})$ and $(n_{c3} - n_{c1})$ are also known.

Equations (73a) and (73b) can be solved by the method of successive approximations: various values are assumed for t_M and plotted against the corresponding calculated m values. The common solution (the point where the two plots cross) gives the proper t_M value.

---

[52] H. J. Gold, *Anal. Chem.* **34**:174 (1962).

Example 4.

Let us calculate the value of t_M for the chromatogram mentioned in Example 3. For the calculation, the values of 90, 95, 100, 105, and 110 were assumed for t_M:

(1) $n_{c2} - n_{c1} = 6 - 5 = 1$

$$t_M = 90 \qquad m = \frac{1}{\log \dfrac{127 - 90}{112 - 90}} = \frac{1}{\log 1.682} = 4.428$$

similarly:

$$
\begin{aligned}
t_M = \ & 95 & m = \ & 3.641 \\
& 100 & & 2.839 \\
& 105 & & 2.011 \\
& 110 & & 1.076
\end{aligned}
$$

(2) $n_{c3} - n_{c1} = 2$

$$
\begin{aligned}
t_M = \ & 90 & m = \ & 3.714 \\
& 95 & & 3.222 \\
& 100 & & 2.701 \\
& 105 & & 2.127 \\
& 110 & & 1.382
\end{aligned}
$$

Figure 79 gives the plots of the two equations for the different t_M values. The point where the two plots cross is 102.8 mm. The agreement of the t_M values is very good:

$$
\begin{aligned}
t_M: \ & \text{measured: } 102 \text{ mm} \\
& \text{Example 3: } 102.6 \text{ mm} \\
& \text{Example 4: } 102.8 \text{ mm}
\end{aligned}
$$

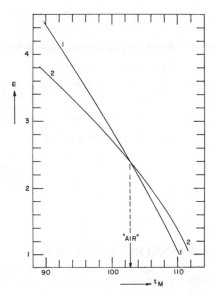

Figure 79. Calculation of the air peak time for Example 4, using the method of Gold.

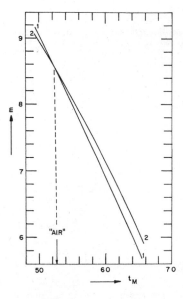

Figure 80. Calculation of the air peak time for Example 5, using the method of Gold.

Example 5.

Let us calculate the t_M value for the chromatogram shown in Figure 6. Three homologous fatty acid methyl esters are selected at random from the sample components:

Component	Carbon number	t_R (mm)
methyl caprylate	11	73
methyl palmitate	17	156
methyl stearate	19	230

For the calculation, the values of 50, 55, 60, and 65 were assumed for t_M:

(1) $n_{c2} - n_{c1} = 17 - 11 = 6$

$$t_M = 50 \qquad m = 9.041$$
$$ 55 \qquad 8.010$$
$$ 60 \qquad 6.910$$
$$ 65 \qquad 5.682$$

(2) $n_{c3} - n_{c1} = 19 - 11 = 8$

$$t_M = 50 \qquad m = 8.953$$
$$ 55 \qquad 8.100$$
$$ 60 \qquad 7.165$$
$$ 65 \qquad 6.086$$

Figure 80 gives the plots of the two equations for the different t_M values. The point where the two plots cross is 52.6 mm.

6.2 INSTRUMENTATION

In the figure captions, the instrumentation used in the analysis is generally given only if its knowledge is essential for the evaluation of the chromatogram. Below, the various instruments, detectors, and readout systems used for the analyses illustrated by the chromatograms are summarized. The commercial gas chromatograms are listed under the name of the manufacturer; the detectors and recorders are characterized by using the following abbreviations:

Detectors
THD—thermistor detector
HWD—hot-wire detector
MTHD—micro thermistor detector
MHWD—micro hot-wire detector
FID—flame ionization detector
DFID—differential flame ionization detector
AID—argon ionization detector

Recorders
PR—potentiometer recorders
GR—galvanometer recorders
OR—oscilloscope readout.

The potentiometer recorders have generally 2.5 or 5 mv full-scale response and a pen speed of 0.25 or 1.0 sec full scale. The galvanometer recorders used are of the 5 ma type, with a full-scale pen speed of 0.1 sec.

Home-made instruments
  with MTHD and OR—Figure 19;
  with MHWD and GR—Figure 20;
  with MTHD and PR—Figures 53, 56, 76;
  with FID and PR—Figures 28, 50, 51, 52, 57, 58, 59.

Barber–Colman Corp. (Rockford, Ill., U.S.A.)
  Model 20 with AID and PR—Figures 30, 32.

Bodenseewerk Perkin–Elmer & Co. (Überlingen, Germany)
  Model 116E with FID and PR—Figures 18, 24, 25, 29;
  Model F/6 with FID and PR—Figures 26, 27, 36, 45, 47, 65, 67.

Carlo Erba Co. (Milano, Italy)
  Model C with FID and PR—Figure 73.

F & M Scientific Corp. (Avondale, Pa., U.S.A.)
  Model 500 with HWD and PR—Figure 38.

Perkin-Elmer Ltd. (Beaconsfield, Bucks., England)
  Model 451 with FID and PR—Figure 31.

Perkin–Elmer Corp. (Norwalk, Conn., U.S.A.)
  Model 154–D with THD and PR—Figures 7, 37;
  Model 154–D with FID and PR—Figures 5, 16, 17, 42, 43, 48, 49;
  Model 154–D with FID and GR—Figure 6;
  Model 800 with FID and PR—Figure 55;

Model 800 with DFID and PR—Figures 74, 75;
Model 800 with FID and GR—Figures 21, 70;
Model 226 with FID and PR—Figures 13, 14, 27, 33, 35, 44, 46, 64, 68, 71, 72;
Model 226 with FID and GR—Figure 66.

Unspecified instrument
  with FID—Figure 54.

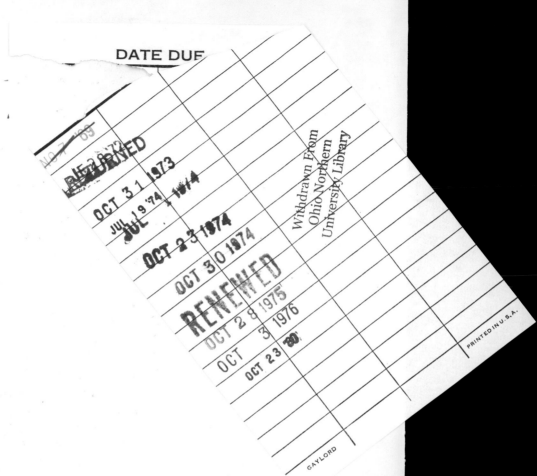